The Politics of Nuclear

The Politics of Nuclear Defence

A Comprehensive Introduction

Greville Rumble

Polity Press

First published 1985 by
Polity Press, Cambridge, in association with Basil Blackwell, Oxford.

Editorial Office: Polity Press, Dales Brewery, Gwydir Street, Cambridge, CB1 2LJ, UK.

Basil Blackwell Ltd
108, Cowley Road, Oxford, OX4 1JF, UK.

Basil Blackwell Inc.
432 Park Avenue South, Suite 1505, New York, NY 10016, USA.

British Library Cataloguing in Publication Data

Rumble, Greville
The politics of nuclear defence: a comprehensive introduction.
1. Nuclear nonproliferation—History
2. Military policy—History—20th century
I. Title
335'.0217 JX1974.73

ISBN 0–7456–0194–4
ISBN 0–7456–0195–2 Pbk

Library of Congress Cataloging in Publication Data

Rumble, Greville.
The politics of nuclear defence.

Bibliography: p.
Includes index.
1. Nuclear warfare. 2. Nuclear weapons.
3. Military policy. 4. Great Britain--Military policy.
I. Title. II. Title: Nuclear defence.
U263.R86 1985 355'.0217 85–18643
ISBN 0–7456–0194–4
ISBN 0–7456–0195–2 (pbk.)

Typeset by Photo·Graphics
Printed in Great Britain by Billing & Sons Ltd, Worcester

Contents

Preface and Acknowledgements

In writing this book I have tried to fairly reflect the arguments put forward both by those who oppose nuclear weapons and those who believe in their utility. I hope that I have provided a comprehensive introduction to the debate on nuclear weapons while avoiding the temptation to thrust my own views forward too forcibly. Inevitably I have drawn on the works of other writers in an attempt to reflect the wide range of arguments put forward on the various issues related to nuclear weapons. While I have had to simplify arguments and to quote people outside the context within which their ideas were first put forward, I hope that I have managed to avoid misrepresenting their views.

I owe a special debt of thanks to Owen Greene who commented on my manuscript; to Anne Rumble who not only read the manuscript in detail in its many drafts but also typed much of it; and to David Held, Jane Katjavivi, Helen Pilgrim and other staff of Polity Press and Basil Blackwell who helped in its final preparation.

Greville Rumble, Woburn Sands, Buckinghamshire

The author and publisher are grateful to the following individuals, publishers and organizations for their permission to reproduce previously published material in this book:

Scientists Against Nuclear Arms (SANA) for tables 2.6 and 2.7, originally published in *The Nuclear Balance*.

Peter Bennett, Malcolm Dando and Castle House Publications for figures 5.2 and 5.3, based on figures originally published in *Nuclear Deterrence: Implications and Policy Options for the 1980s* (edited by Barrie Newman and Malcolm Dando).

Stan Openshaw, Philip Steadman, Owen Greene and Basil Blackwell Ltd for tables 6.1, 6.2 and 6.3, based on tables originally published in *Doomsday: Britain after Nuclear Attack*.

Julie Dahlitz and George Allen and Unwin Ltd for material from *Nuclear Arms Control* incorporated into chapter 8.

1
The Start of the Atomic Age

In 1935 Leo Szilard, a Hungarian physicist, faced with the rise of fascism in Germany, resigned his post at the University of Berlin and moved to London. Two years previously he had come to believe in the possibility of building an atomic bomb and was convinced that Nazi Germany would attempt to develop one. He failed to impress the British Army with his theoretical super-weapon but the Royal Navy took a greater interest and in 1936 he took out two British patents, both classified Top Secret, suggesting ways of making an atomic bomb. By 1938, Szilard's theories were being confirmed. Two physicists at the University of Berlin, Otto Hahn and Fritz Strassmann, and the French Nobel prize winner Frederic Joliot-Curie, conducted experiments which supported Szilard's views that uranium atoms could be split and that a chain reaction was possible. As Europe moved closer to war physicists in France, Germany, and the USSR began to work on the development of an atomic bomb. Szilard himself moved to the USA, where in collaboration with an Italian refugee, Enrico Fermi, he tried unsuccessfully to persuade the US Navy to take an interest in their research.

In Britain and France fear that the Germans were at work on a bomb led both governments to start research projects but belief in the possibility of making an atomic bomb was strained because of the amount of uranium ore needed to start a chain reaction. Only when two refugee German research physicists, Otto Frisch and Rudolph Peierls, suggested that it might be possible to extract pure uranium-235 from uranium ore did it begin to seem possible. Their paper, known as the Frisch–Peierls Memorandum, persuaded the British government to finance a bomb project (code-named the Maud Committee). The British, however, came to the conclusion that they could not support research into an atomic bomb and, during 1941, they sent the Maud Committee report entitled 'The Use of Uranium for a Bomb', to the Americans.

Szilard failed to interest the US Navy in his project but persuaded Albert Einstein to write to President Roosevelt and try to interest him in the project. Einstein's letter of 2 August 1939 had the desired effect: Roosevelt established the Uranium Committee to study the prospects of a bomb. The British Maud Committee report gave added impetus to American initiatives, and led to the establishment of the Manhattan Project under Major-General Leslie Groves. His task was to make the world's first atomic bomb.

Between the autumn of 1942 and the end of 1944 Groves established a new industry. In the process, enormous plants were built: 'the first fully automated factory, the first plant completely operated by remote control, the first totally leakproof industrial system' (Pringle and Spigelman: 1982, p. 26). By the spring of 1945 the scientists and engineers involved in the project were ready to begin assembling a bomb. The collapse of Germany and the realization that the Nazis had no bomb led some of the scientists to question, without avail, the use of the bomb on moral grounds. General Groves was determined to continue. On the eve of Germany's surrender he sent out an order in which he said that the project 'will continue and increase after VE day with Japan as the objective' (Groves, 1945).

The decision to actually use the bomb against Japan was taken by President Truman in June 1945. The precise nature of the atomic attack was a matter of some discussion. Towards the end of May, a demonstration explosion was proposed and discussed by the presidential committee which had been established earlier that month to advise on a future policy for atomic energy. Called the 'Interim Committee' by its Chairman, Secretary of War Henry L. Stimson (who believed that Congress would wish to appoint a permanent committee at a later date to supervize, regulate and control the development of atomic energy), it agreed that while the USA 'could not concentrate on a civilian area', it should 'seek to make a profound psychological impression on as many of the inhabitants of Japan as possible'. The minutes of the committee record that Stimson 'agreed that the most desirable target would be a vital war plant employing a large number of workers and closely surrounded by workers' houses' (Minutes of the Committee, cited in Sherwin, 1975, p. 302).

Six of the Manhattan scientists formed a 'Committee on Political and Social Problems' under the chairmanship of James Franck and proposed yet again that a demonstration bomb be dropped on an uninhabited place where the Japanese could appreciate the nature of the weapon without experiencing its force. Killing thousands of civilians with the new bomb would, they argued, damage America's post-war image and undermine efforts to reach an international agreement on the control of the atom. Their report was considered by the Scientific Panel of the Interim Committee but the panel decided that use of the bomb would bring the war to an earlier end and save other, American, lives. They therefore advised Truman that there was no alternative to the military use of the atomic bomb.

Some historians, for example, David Horowitz (1967, pp. 53–5) have placed the decision to use the bomb in the wider context of relationships between the USA and the Soviet Union, and in particular the Truman Administration's wish to exclude the Soviets from any settlement in the Far East. According to this version of events, as the war in Europe drew to a close the Allies began to consider the nature of the international post-war settlement. In February 1945 the 'Big Three' – Churchill, Roosevelt and Stalin – met in Yalta to consider the political future of Europe and the question of Soviet entry into the war against Japan.

The Soviet Union had borne the major weight of the war against Germany since June 1941. In the process a large part of the country had been overrun and millions of its citizens killed. (As a measure of comparison, 20 million Soviet citizens were killed during the Second World War while American military losses amounted to no more than 405,000). It was not, therefore, surprising that the Soviet Union did not immediately enter the war against Japan although as early as 1943 Stalin promised that he would do so eventually. At Yalta, however, Stalin pledged himself to declare war on Japan three months after the end of hostilities in Europe. At this time neither Roosevelt nor Churchill were confident of their ability to defeat Japan easily without Soviet help. With the outcome of the Manhattan Project (which had been kept secret from the USSR) still uncertain, the British and Americans agreed to guarantee Soviet rights of access to the Manchurian ports of Dairen (now Talien), which was to be an international port, and Port Arthur (now Lushun), which was to be leased to them as a naval base. In return the Soviets agreed that the Allies should treat with Chiang Kaishek's regime in China, rather than with the communists under Mao Zedong.

On 8 May 1945 Germany surrendered, thus committing the USSR to enter the war in the Far East on 8 August 1945. Shortly after, the Big Three, with Truman in the place of Roosevelt, met at Potsdam on 17 July. The day before, at 5.30 a.m., the successful test firing of an atomic bomb had taken place at Alamogorda in New Mexico.

The success of this test led to a radical change in US thinking. While the bomb remained primarily a weapon to be used to end the war in the Pacific as quickly as possible, it might also end it before, as US Secretary of State James Byrnes put it, the USSR could 'get in so much on the kill' (Yergin, 1980, p. 116). The USA and Britain, in conjunction with China, hurriedly issued an ultimatum to the Japanese and called on them to surrender unconditionally or face 'prompt and utter destruction' (26 July). No prior warning of the ultimatum was given to Stalin although the Potsdam Conference was still in progress, nor was any reference made to the fact that, as early as 13 July, the USA had learnt through intercepted messages from Togo to the Japanese Ambassador in Moscow that the Japanese were interested in sueing for peace though not on the basis of unconditional surrender. Peace negotiations could hardly have been concluded before 8 August, when the USSR was due to enter the war. From the British and American point of view the emphasis was on the 'prompt' surrender of Japan. The US Navy Secretary James Forrestal noted in his diary on 28 July that US Secretary of State Byrnes had 'said he was most anxious to get the Japanese affair over with before the Russians get in' and that in particular he did not wish the USSR to get Dairen and Port Arthur (Forrestal, 1951). Only immediate and unconditional Japanese surrender would have ended the war quickly. The Japanese failure to respond to the Potsdam Proclamation calling for unconditional surrender sealed the fate of Hiroshima and Nagasaki.

On 6 August the first atomic bomb was dropped on Hiroshima. On 8 August, true to Stalin's promise, the USSR declared war on Japan. On 9 August the second bomb was dropped on Nagasaki.

On the 10th, Japan made clear its intention to surrender and on the 14th it did so unconditionally.

The news of the successful atomic bomb test at Alamogorda was recognized by Western leaders as a quantum leap in the technology of warfare. Byrnes commented on 29 July that 'the New Mexico situation has given us great power' (Davies, diary entry for 29 July 1945). Alan Brooke, the British Commander-in-Chief, in his diary entry for 23 July, noted that Churchill was 'carried away' by the news. 'We now have something in our hands which would redress the balance with the Russians', Churchill said to him. 'The secret of this explosion and the power to use it would completely alter the diplomatic equilibrium which was adrift since the defeat of Germany' (Bryant, 1959, pp. 363–4). Stalin's reaction was to accelerate the Soviet Union's own bomb programme.

References and Further Reading

Bryant, A. (1959) *Triumph in the West. Based on the Personal Diaries of Field Marshal Lord Alanbrooke*, Garden City: Doubleday.

Davies, J. Unpublished Diaries, Library of Congress. Cited in D. Yergin (1980) *Shattered Peace. The Origins of the Cold War and the National Security State*, Harmondsworth: Penguin Books.

Forrestal, J. V. (1951) *The Forrestal Diaries*, ed. W. Millis, New York: Viking Press.

Groves, L. (1945) Memorandum, 8 May, Ernest Lawrence papers, Carton 30, Folder 23, cited in P. Pringle and J. Spigelman (1982) *The Nuclear Barons*, London: Sphere Books Ltd., p.34.

Horowitz, D. (1967) *From Yalta to Vietnam. American Foreign Policy in the Cold War*, Harmondsworth: Penguin Books.

Pringle, P. and Spigelman, J. (1982) *The Nuclear Barons*, London: Sphere Books Ltd.

Sherwin, M. (1975) *A World Destroyed. The Atomic Bomb and the Grand Alliance*, New York: Knopf.

Yergin, D. (1980) *Shattered Peace. The Origins of the Cold War and the National Security State*, Harmondsworth: Penguin Books.

2
Nuclear Weapons Since 1945

The Failure of International Efforts to Control the Bomb, 1945–9

American policy makers had begun to consider how the atomic bomb might be used as an element in US foreign policy even before the defeat of Japan and at the first post-war Council of Foreign Ministers, convened in London in September 1945, the US Secretary of States, Byrnes, was optimistic. He believed that the atomic bomb would strengthen his hand in the negotiations on the future of Eastern Europe. The US Secretary of War, Henry Stimson, commented that Byrnes, then preparing for the conference, 'looked to have the presence of the bomb in his hip pocket, so to speak' (Stimson, 1945). Stimson, in contrast, rejected the view that the US should capitalize on the atomic bomb secret. He warned President Truman that the Soviet government would regard the US acquisition of the bomb as a threat and that 'the temptation will be strong for the Soviet political and military leaders to acquire this weapon in the shortest possible time' (Stimson and Bundy, 1947, pp. 642–6). He proposed that the USA should share its scientific information with the USSR.

Stimson's ideas were discussed at a US Cabinet meeting on 21 September. Truman asked members to submit their views in writing. The Under-Secretary of State, Dean Acheson, in his response, warned that 'any long-range understanding based on firmness and frankness and mutual recognition of the other's basic interests seems to me impossible under a policy of Anglo-American exclusion of Russia from atomic development. If it is impossible, there will be no organised peace but only an armed truce' (Yergin, 1980, p. 133).

In evaluating the impact of the atomic bomb on foreign policy objectives US leaders were faced with the problem of assessing how long the US monopoly would last. The engineers and scientists who had developed the bomb believed that the Soviet Union would have the capability to make one within three to five years. Soviet scientists already had the scientific knowledge; their problem was essentially an engineering one. It was therefore reasonable to suppose that the Soviet Union would try to catch up with the USA. US military advisers believed, however, that the American monopoly would last for many years, if not decades. Vannevar Bush, chief scientific adviser to the President, was to comment later of Admiral Leahy that 'his view was... [that] there was an atomic "secret" ... some sort of magic formula. If we guarded this, we alone could have atomic bombs indefinitely' (Bush, 1970, p. 295). Major-General Groves, the head of the Manhattan Project, suggested that the USSR would need at least 20 years to develop a bomb. This view was shared by President Truman.

Truman did not initially rule out some form of world-wide control of atomic weapons. In December 1945 Secretary of State Byrnes, visiting Moscow, obtained Soviet agreement to discuss the issue at the newly established United Nations. Accordingly, at the first meeting of the UN General Assembly, convened in London on 24 January 1946, the first resolution was to establish an Atomic Energy Commission charged among other things with producing proposals 'for the elimination from national armaments of atomic weapons'. The resolution was unanimously adopted by all 51 member states.

The US government had, meanwhile, anticipated events by establishing a consultative panel under the chairmanship of David Lilienthal to consider proposals for the establishment of a UN Commission to control the international development of atomic energy. The panel began by distinguishing between the peaceful and military uses of atomic energy, and proposed the establishment of an International Atomic Development Authority which would control its 'dangerous' development by owning or controlling all uranium mines and all plants for the production of fissile material. Only small-scale research reactors incapable of producing enough fissile material for a bomb would be uncontrolled. While there were problems, particularly in respect of verification of mining and production within the Soviet Union, these proposals, known as the Acheson–Lilienthal proposals, had the full support of the panel. But Byrnes, to the consternation of Acheson and Lilienthal, then appointed the conservative Bernard Baruch as the US representative to present the proposals to the United Nations.

In the heightening tensions of the early Cold War period (Churchill's 'Iron Curtain' speech was delivered on 5 March) Baruch established his own advisory committee to review the Acheson–Lilienthal proposals. He also consulted the military. Army Chief of Staff Dwight D. Eisenhower wrote that 'we cannot at this time limit our capability to produce or use the weapon' (Yergin, 1980, p. 238). The military's position was basically that 'the bomb should continue to be at the heart of America's arsenal, and a system of controls should be established that would prevent the Russians from developing the weapon' (Yergin, 1980, p. 238). As a result of his discussions, Baruch also came to the conclusion that the Russians would not be able to develop a weapon or tap uranium supplies for many years.

On 14 June 1946 Baruch presented his proposals to the UN. Under these, the manufacture of atomic bombs would have been prohibited and all phases of the development and use of atomic energy placed under an international authority. Compliance was to be assured through verification. However, until such time as a UN treaty could be agreed upon and come into force, the USA was to retain its monopoly on atomic secrets and did not promise to end production of new atomic weapons. The effect of this plan would have permitted the USA to continue to build and develop nuclear weapons until such time as there were effective and foolproof international controls, including on-the-spot verification within Soviet territory. It also envisaged the 'swift and sure punishment' of those states violating the treaty. Since the USA was the only state capable of carrying out such punishment and the Soviet Union the only likely state at the time to be trying to develop atomic weapons, the proposal 'could be interpreted by Moscow as an attempt to turn the United Nations into an alliance to support a United States threat of war against the USSR unless it ceased its efforts [to make a bomb]' (Acheson, 1970, p. 155).

Soviet Foreign Minister Andrei Gromyko rejected the American proposals and on 19 June submitted an alternative which would have bound all parties not to use atomic weapons; forbidden the production or keeping of any such weapons; and required the destruction of all weapons then in existence within three months of the treaty coming into force. The Soviet Union offered no means of verification in support of their proposals, which made them wholly unacceptable to the USA. There is some evidence that the US negotiators knew that insistence on on-the-spot inspection would make their proposals unacceptable to the Soviet Union. The US then made absolutely sure that the search for international controls would fail by exploding the world's fourth atomic bomb over Bikini Atoll on 1 July 1946.

The immediate Soviet reaction to the Bikini Test was to accuse the USA of trying to influence the talks. The US action, said *Pravda* 'fundamentally undermined the belief in the seriousness of American talk about disarmament' (Pringle and Spigelman, 1983, p. 55). On 5 July the USA formally rejected the Soviet proposals, and although Baruch forced a vote on his own proposals through the UN Security Council on 31 December 1946 the Russian objection and the failure of the negotiations in the previous June effectively ended this first attempt to control atomic weapons. The resulting impasse lasted until 1950.

The Age of the Bomber

Evaluation of the first post-war bomb tests by the United States Air Force (USAF) confirmed the power of the weapon and thrust the Air Force into the nation's first line of defence. In March 1946 the Strategic Air Command (SAC) was established. The Cold War, and particularly the Berlin Crisis of 1948, led the Truman Administration to consider for the first time the need to be prepared to deliver US atomic weapons against the Soviet homeland (Rosenberg, 1983, pp. 12–13). Under General Curtis LeMay SAC was expanded and began to plan for the atomic bombing of Soviet cities (see table 2.1).

USAF war plans were used to justify increased weapons production, which in turn justified increased appropriations to provide aircraft to deliver the bombs. The stockpile of bombs generally lagged behind the requirements laid down in the annual war plans. Not until the early 1950s did the first mass produced atomic bomb, the Mark 4, begin to enter service. By 1952 the problem of supply was essentially solved and from then on the US stockpile began to rise rapidly. The number of nuclear capable aircraft at SAC's disposal also began to rise. Initially SAC had only 30 modified

Table 2.1 Early US war plans

Date	War plan	No. of bombs	No. of targets	Type of target
1947	Broiler	34	24	Cities
1948	Halfmoon	50	20	Cities
1948	Trojan	133	70	Cities
1949	Offtackle	220	104	Cities
		72	—	Reserved

B-29 bombers capable of carrying the Mark 3 atomic bomb. By December 1948 this had risen to 60 aircraft. In June of that year it received its first B-36 long-range intercontinental bomber. Although the replacement for the B-36, the B-52 heavy jet bomber, was already on the drawing board, the USAF gave priority to its new B-47 medium-range jet bomber, the first prototype of which flew in December 1947. In part this was because the technical problems faced were less great but it also reflected the fact that the USA had been acquiring forward-bases in Europe since 1946. In 1948 the USAF stationed six squadrons of B-29s in Britain and in 1953 it began to station B-47s there.

On 29 August 1949 the Soviet Union exploded its first atomic device. Evidence of the explosion shocked US leaders and led Truman's three-man advisory committee on nuclear policy to endorse the proposed acceleration of the atomic energy program put forward by the military on the grounds that it was necessary in the vital interests of national security. Among the projects approved by Truman, although not without considerable debate and against the advice of the Chairman and a number of members of the US Atomic Energy Commission's General Advisory Committee, was the H-bomb. Thermonuclear devices were tested in May 1951 and November 1952. However, it was not until 1 March 1954 that the US successfully tested a usable H-bomb. Twenty months later (November 1955) the Soviet Union tested an H-bomb of its own.

When President Eisenhower entered office in 1953, the US military budget was swollen by the costs of the Korean War. Convinced of the need to reduce expenditure, Eisenhower's 'New

Look' defence policy emphasized nuclear weapons not just as weapons of war but as a means of deterring communist expansion in Europe and Asia, thus allowing the USA to support its world-wide security interests without having to match communist manpower in each theatre. Reflecting these priorities, the Administration abandoned the objective, outlined in National Security Council document NSC-68/4 (December 1950), of a phased and balanced build-up of American forces. During 1951 priority was given to the Air Force. By the end of 1953 SAC had over 1,500 aircraft including 1,000 nuclear-capable bombers, a significant increase on the 250 nuclear-capable aircraft of June 1950. The propellor-driven B-29s were being phased out in favour of the B-47, and in June 1955 the first B-52 heavy intercontinental jet-bombers came into service. By 1959 they had completely replaced the propellor-driven B-36s. In that year SAC reached its peak capability with nearly 500 B-52s, more than 2,500 B-47s and over 1,000 propellor and jet-driven tanker aircraft. Development of the supersonic B-58 (which entered service in August 1960) and of the proposed B-70 'Valkyrie' follow-on bomber was also under way.

As well as its forward-bases in Britain, SAC acquired airfields in Libya, French Morocco and Spain, and had the right to move bomber groups into Iceland, Cyprus and Turkey in time of crisis. In the Pacific, it had bases in Guam, Hawaii and the Philippines. SAC's need for such bases was reduced somewhat with the introduction of the intercontinental B-52 bomber and some of its bases in Britain were transferred to the USAF Tactical Air Command whose nuclear-capable fighter-bombers were expected to support ground operations in a European war. By the mid-1960s three of SAC's four remaining UK bases had been handed back to the RAF and its bases in Morocco closed.

The Soviet Union never attempted to match the USAF's bomber capability. The Soviet Tupolev Tu-4 Bull medium-range bomber, a copy of the US B-29, first appeared in May 1947. New aircraft types were introduced in the 1950s, including the Tupolev Tu-16 Badger medium-range bomber (1954) and the Myasischev M-4 turbojet Bison, which had a range of 10,000 km and was first demonstrated at the 1954 May Day parade in Moscow. US intelligence grossly over-estimated and over-projected the number of M-4 bombers deployed. As a result, USAF Intelligence estimated that by mid-1959 the USSR would have a force of from 600 to 700 intercontinental bombers. While other US intelligence agencies were more sceptical and suggested a fleet of about 500 aircraft, these figures were used to raise American fears about a 'bomber gap' and to justify the procurement of another 538 B-52s.

At the beginning of 1961 the USSR had about 190 long-range bombers and at the height of their deployment, perhaps 210. The fear that the Soviet Union might nevertheless develop such a force led the USA to develop air defences, including the establishment by 1953 of a continuous 150-mile-range radar net throughout the USA, and the extension of this network beyond US frontiers with the construction of the Distance Early Warning (DEW) Line from Alaska across Canada into Greenland. The main section of this line was completed by 1957 and would have provided 5 to 8 hours' warning of a Soviet bomber attack. Linked with these preparations was the deployment of interceptor squadrons within the USA and the development and deployment of anti-aircraft missiles. By 1957, however, these programmes were being cut back in the face of the much more serious threat posed by the development of missiles.

The Early Development of ICBMs

During the Second World War the Germans had begun to develop and deploy both cruise missiles (the V-1) and ballistic missiles (the V-2). With the collapse of Germany both the USA and the Soviet Union acquired copies of the V-2s. By the end of 1945 the US had a rocket research establishment at White Sands in Nevada and by March 1946 had begun test-firing V-2s. In 1954 the Strategic Missile Evaluation Committee reported that ballistic missile technology had reached a stage at which the development of an intercontinental ballistic missile (ICBM) seemed feasible, a conclusion endorsed by a Rand Corporation study. As a result the USAF accelerated development of a liquid-fuelled ICBM (the Atlas) and in 1955 the National Security Council gave top priority to this programme. In the same year the USAF began to develop the more advanced Titan ICBM and started research on solid-propellant rockets in the intercontinental range – a project that eventually led to the approval (in 1958) of the Minuteman ICBM programme. Consideration was also given to protecting these ICBMs in hardened launch facilities and concrete silos.

Soviet scientists had undertaken pioneering work on rocket engineering during the 1930s. After Stalin's purge (1936–8), in which many of the leading engineers were arrested, research was confined to the development of rocket artillery, but with the appearance of V-1s and V-2s interest was rekindled and by 1947 Soviet scientists had tested a modified version of the V-2. By 1950, work had begun on the development of the SS-3, a medium-range missile. On 4 October 1957 the Soviet Union launched the first satellite into space on the back of an SS-6. While the SS-6 was of limited military use (only four were ever deployed), the launch proved that the Soviet Union was capable of building intercontinental-range ballistic missiles that could reach the USA. A month later the Gaither Committee (November 1957) warned that 'by 1959, the USSR may be able to launch an attack with ICBMs carrying megaton warheads, against which SAC will be almost completely vulnerable under present programs' (Security Resources Panel, 1957, p. 14).

The Gaither Report recommended that immediate steps should be taken to protect SAC's bombers from surprise attack. In January 1958 approval was given for the construction of the Ballistic Missile Early Warning System (BMEWS)to complement the DEW Line. Development also began on the 'Midas' early warning satellite and significant progress was made on improving SAC's response time. By 1959 one-third of the bomber force was being held on ground alert at any one time and plans were laid for part of the bomber force to be held on airborne alert, flying along Emergency War Plan routes, their missions being aborted unless they received the 'go' code to continue. More importantly, the USA began to accelerate its own missile programmes. In September 1959 the first Atlas ICBMs were brought into service, while the first Titan ICBM was successfully tested during 1960. The fact that relatively few of these liquid-fuelled missiles were built was due to the success of the Minuteman solid-fuelled ICBM and Polaris submarine-launched ballistic missile programmes.

Theatre Nuclear Weapons

During 1950 and 1951 the USAF had begun to deploy relatively short-range fighter-bomber squadrons operating F-100 Super Sabre and F-101 Voodoo nuclear capable aircraft from bases in Britain, together with B-66 Destroyer medium-range bombers. In the Mediterranean, the non-nuclear components of atomic bombs were deployed aboard the USS *Franklin Roosevelt* during 1950. Subsequently the US Navy deployed AJ-1 carrier-borne Savage aircraft armed with atomic bombs in the Mediterranean. Tactical nuclear weapons also began to be deployed by US ground troops in Europe. In all, some 7,000 such weapons were brought to Europe during 1953–4, including the short-range (40 km) Honest John low-kiloton yield, unguided battlefield missile (from 1954), the 800 km Matador cruise missile (from 1954) and the Mace A/B 2,500-km-range cruise missile (from 1956).

In 1953 the US National Security Council authorized the Joint Chiefs of Staff to develop plans for the deployment of Intermediate Range Ballistic Missiles (IRBMs). The decision to develop them was taken in 1955 following receipt of the Killian Committee report 'Meeting the Threat of Surprise Attack'. By the end of the year the USAF was developing the Thor IRBM for deployment in Western Europe, while the US army began to develop the liquid-fuelled Jupiter IRBM. The Thor, with a range of 3,000 km and a warhead yield of 1.5 megatons, was deployed in Britain by the USAF in 1958. Subsequently 60 Thor missiles were installed in Britain under 'dual-key' control (December 1959), while 30 Jupiter IRBMs were deployed under similar conditions in Italy and 30 in Turkey (in 1961). The liquid-fuelled Thor missiles operated from fixed sites and while they were protected by shelters both they and the Jupiters were, by the early 1960s, vulnerable to attack. They were also not particularly accurate although to some extent this was compensated for by their yield (1.5 megatons). As a result the Thors were withdrawn in 1963 and the Jupiters in 1965. The USA's theatre nuclear capability came to be based on the Polaris SLBM and also on the Pershing I, which was first deployed in Europe in October 1960, and on nuclear-capable F-4 and F-104 aircraft which were introduced during 1959–63.

The deployment of Soviet IRBMs, which began in 1957 with the SS-3 followed by the SS-4 (2,000-km-range) in 1959 and SS-5 (3,500-km-range) in 1961, threatened Western Europe. The SS-4s and SS-5s operated from fixed locations and took from 8 to 24 hours to prepare for firing. Very few of them were silo-based and they, like the US Thor missiles in Europe, were judged to be very vulnerable to surprise attack. As a result the Soviet Union began to withdraw the SS-4s and SS-5s, replacing them with SS-11

ICBMs adapted for use against European targets. Some of the SS-4s and SS-5s were redeployed in the Far East where, by the mid-1960s, Soviet relations with China had deteriorated.

At tactical (battlefield) level US and NATO forces came to rely increasingly on nuclear weapons. The US Department of Defense began to study the use of tactical nuclear weapons in 1951 (the 'Project Vista' study). The military arguments favouring their deployment stressed the saving in US and NATO conventional force levels, the fact that their use would discourage the Soviet Union from massing its ground forces to attack the West, and the fact that they made 'forward defence' (that is, defence at the West/East German frontier) feasible. The concept of using battlefield weapons to defend West Germany was shown to be ridiculous when a 1956 military exercise ('Carte Blanche'), simulating the use of 355 nuclear weapons in Germany by NATO, resulted in enormous casualties (1.7 million Germans dead and 3.5 million wounded).

ICBMs and SLBMs

The closing months of the Eisenhower Administration were marked by fears, which the Soviet leader Khruschev did his best to sustain, of a 4 : 1 Soviet lead in ICBMs by the early 1960s. Senator Kennedy, during his election campaign in 1960, laid stress on the supposed weakness of the USA. He came to the Presidency committed to a significant build-up in US conventional and nuclear forces. The 'missile gap' was, however, quickly shown to be an illusion.

Ever since the early 1950s SAC had carried out limited overflights along the borders of the Soviet Union, both to test Soviet air defences and to spy. In 1956 the US began to send high-flying aerial reconnaissance missions deep into Soviet territory using U-2 aircraft flying out of bases in England, Turkey and Japan. Some 20 flights took place between 1956 and 1958. By 1959 the Russians had begun to try to shoot the aircraft down, and the flights were suspended temporarily. Then in May 1960 a U-2 flying from Peshawar in Pakistan to Norway was shot down, bringing the flights to an end. By then Lockheed had begun to develop photographic reconnaissance satellites on behalf of the USAF. The first successful launch of a Discoverer satellite (Discoverer 14) took place in August 1960, the satellite circling three times over the Soviet Union, taking photographs, before falling back to earth where it was recovered. Between 1959 and 1962 38 Discoverer satellites were launched. The USA also began to develop better ways of sending information back to Earth using radio relays in SAMOS satellites, the first successful launch of which took place in January 1961.

As a result of this intelligence gathering US estimates of Soviet missile strength were steadily revised downwards during 1961 to 3.5 per cent of their original December 1959 size. From a lead of 18 ICBMs to 4 in 1960, the USA had, by 1964, established a commanding advantage over the Soviet Union of 834 ICBMs to 200. By the mid-1960s the USA had a highly effective and survivable (i.e. silo and submarine based) strategic nuclear force comprising 1,054 ICBMs (1,000 Minuteman ICBMs and 54 Titan II ICBMs), 656 SLBMs and 630 B-52 bombers. The first US ballistic missile submarine (SSBN), the USS *George Washington*, was commissioned in December 1959. By 1967 the US navy had 41 SSBNs, each carrying 16 Polaris missiles, and providing a secure second-strike capability.

Once it had gained a strategic superiority, the USA sought to preserve this through proposals for arms control that were initially prepared by the Kennedy Administration and subsequently endorsed by President Johnson. The proposals, which would have frozen existing force levels and forbidden the introduction of significant new types of vehicles, were presented to the Eighteen Nation Disarmament Conference in January 1964, and were promptly rejected by the Soviet Union. A counter proposal by the Soviet Union to reduce each side's missiles to extremely low levels (the Gromyko plan) was rejected by the USA on the grounds that it could not be adequately verified.

By the middle of 1962 Soviet leaders were convinced that they had to respond to the US military build-up. They tried to deploy medium-range and intermediate-range ballistic missiles in Cuba. In the face of strong US reaction, however, they had to abandon this plan. This led directly to Khruschev's decision to accelerate the production of the 11,000 kilometre range SS-7 ICBM, which was first introduced into service in 1962. The following year the first SS-8 ICBMs, which also had a range of 11,000 km, were deployed. At the same time, development began on the more advanced SS-9 and SS-11 ICBMs which could be fired from 'hardened' (i.e. protected) launch pads or, in the case of the SS-11, from silos. Development of a Polaris-type SLBM was also authorized.

Khruschev nevertheless refused to authorize the massive missile programme that some of his colleagues called for. After his fall from power (1964) this policy of restraint was abandoned. The new Kremlin leaders, Leonid Brezhnev and Alexei Kosygin, pressed on with the build-up of Soviet strategic forces. SS-11s began to be deployed during 1966, followed by SS-9s towards the end of 1967. As a direct result the size of the Soviet ICBM force increased from around 200 in 1964 to 340 in 1966 and 730 in 1967. By the end of 1968 the Soviet Union was approaching parity with the USA in the number of hardened ICBMs deployed and had also begun to deploy operational SLBM submarines.

MRVs and MIRVs

During 1961 intelligence reports began to circulate suggesting that the Soviet Union was constructing a string of ABM sites (the Tallinn Line) stretching from Archangel to Riga. There were also reports of an anti-ballistic missile (ABM) system around Leningrad. This was briefly operational in 1962. In 1964 another ABM system (the Galosh) began to be deployed around Moscow. The US response was to develop multiple re-entry vehicles (MRVs) and, a little later, multiple independently targetable re-entry vehicles (MIRVs).

The early missiles had flown a pre-programmed trajectory from a given launch point to a selected target. Their direction and distance were determined during the initial powered flight of the rocket. The motors were shed when the rockets burned out and a re-entry vehicle (RV) then carried on under its momentum towards the target. The new MRVs involved a group of warheads fitted to a re-entry vehicle from which they would separate and then fall towards the target in a cluster formation, thus increasing the area of destruction and also to some extent making up for missile inaccuracy. By 1964, however, improved guidance systems and the development of 'space bus' technology (the 'bus' is the front section of a missile) enabled the US to begin developing MIRVs. In a MIRVed system the warheads are attached to the bus, which is covered by a nose cone. When the last rocket motor burns out and separates, the nose cone is ejected. The bus minus its cone then goes through a long coasting phase, dropping off warheads at predetermined points. Since these can hit different targets, they are said to be 'independently targetable'.

During 1964 the decision to develop the Poseidon C-3 and Minuteman III MIRVed SLBM and ICBM was taken. As the programme gathered momentum it became clear that Soviet ABM capabilities had been greatly exaggerated. Evidence was obtained showing that the Tallinn Line was an anti-aircraft defence system. Moreover, there were no indications that the Soviet Union was developing further Galosh-type ABM systems to protect its cities. In 1968 US intelligence services were admitting as much. However, by then they were suggesting that the Soviet Union also had a MIRV programme and its imminent deployment was predicted for 1969. In fact, although the USSR did test a crude multiple-warhead system that year, it was not until the late 1970s that an effective Soviet MIRVed missile (the SS-18) was deployed.

The increasing size of the Soviet Union's strategic nuclear forces led, however, to American efforts to seek means of defending the USA against Soviet attack. The US Secretary of Defense, Robert McNamara, regarded ABM systems as a waste of money although he had approved an ABM research programme in the early 1960s. In 1967, however, under intense pressure from President Johnson and the Joint Chiefs of Staff, he approved the building of a nation-wide 'light' ABM system (the Sentinel system) designed to protect the USA from a Chinese nuclear attack or an accidental Soviet ICBM launch. The Nixon Administration, which came into office in 1969, modified the Sentinel system, and this became the Safeguard system designed to protect the US Minuteman ICBMs from a Soviet first strike. It was later to be extended to cover the whole nation. Critics of the system argued that it was costly and likely to be, at best, only marginally effective. By 1970 Defense Secretary Melvin Laird was admitting that it could not protect the US missile fields against a fully developed Soviet MIRV capability and Congress eventually rejected Nixon's requests for funds to construct ABM sites.

The Ostensible Achievement of Parity: SALT I and SALT II

Both the USA and the Soviet Union became increasingly conscious as the 1960s wore on that unless they could reach an agreement they would be locked into an expensive arms race. Following informal contacts, President Johnson was able to announce in March 1967 that Soviet leaders were willing to discuss means of limiting the arms race in offensive and defensive nuclear missiles. Progress was slow, however, and it was not until the spring of 1968 that the Soviet Union agreed to begin exploring mutual arms control. By August the USA had drawn up initial proposals but the Soviet invasion of Czechoslovakia led the Administration to cancel a proposed summit meeting. Then in November Johnson lost the presidential election and a hiatus ensued during which the new Nixon Administration studied the various proposals on the table.

Discussions with the Soviet Union resumed in June 1969 and the first round of talks opened in November. The talks culminated in the signing of the ABM Treaty (May 1972) which restricted both superpowers to a total of 200 ABM interceptors to be deployed at two separate sites of 100 launchers each, one at each nation's capital and the other at an ICBM field. At the same time a five-year Interim Agreement on offensive strategic weapons limited land-based missiles to existing levels, placed special restrictions on the number of large ICBMs, and set a ceiling on submarine-based missile systems. Provisions were made for the replacement of existing systems by more modern ones and force levels were temporarily fixed as shown in table 2.2. No restrictions were placed on bombers, forward-based systems, or MIRVs, in all three of which the USA led. The Agreement formally expired on 3 October 1977.

Reactions to the ABM Treaty were generally favourable within the USA, but conservatives argued that the Interim Agreement placed the USA in an inferior position. Its passage through the Senate was therefore formally linked to an amendment calling for further discussions which would not disadvantage the USA. The Agreement and the ABM Treaty were formally implemented by the Soviet Union and the USA in October 1972 and in the following month the next series of talks (SALT II) opened.

Table 2.2 The SALT I interim agreement (26 May 1972)

Missiles	USSR	USA
ICBMs (maximum)	1618	1054
SLBMs	740	656
ICBMs + SLBMs (maximum)	2358	1710
SLBMs on submarines (maximum limits)	950 on 62	710 on 44

Note: Maximum limits on SLBMs could only be reached by phasing out ICBM launchers. The total number of offensive launchers was to remain constant.

During 1973 the USA and the Soviet Union reached agreement on the need to replace the 1972 Interim Agreement by a more permanent one, but in the event it proved impossible to do this quickly, and they therefore entered into yet another interim agreement in November 1974 (the Vladivostok Accord) which established the principle of equal ceilings on strategic delivery vehicles (2,400 ICBMs, SLBMs and heavy bombers) and on the number of launchers (1,320) that could be equipped with MIRVs. Within these limits each side could compose its forces as it wished. The final agreement signed in Vienna on 18 June 1979 included a treaty which imposed limits on strategic nuclear offensive weapons until 31 December 1985 (the SALT II Treaty); a protocol, integral to the treaty, which set forth certain limitations until 31 December 1981 (SALT II Protocol); and a joint statement of principles and basic guidelines for subsequent negotiations on the limitation of strategic arms, leading to SALT III.

Under this Treaty, certain aggregate limits were set for each side within the general framework that had been established at Vladisvostok (see table 2.3). It also imposed a freeze on the number of re-entry vehicles (RVs) on current types of ICBM, with a limit of 10 RVs on each of the one new type of ICBM allowed each side; a limit of 14 RVs on SLBMs; and a limit of 10 RVs on air-to-surface ballistic missiles (ASBMs). ASBMs were also to be included in the aggregate although neither side had plans to deploy any before 1985. An average of 28 long-range air-launched cruise missiles

Table 2.3 The SALT II agreement (18 June 1979)

	Aggregate limit
(a) Launchers of MIRVed ICBMs	820
(b) Launchers of MIRVed ICBMs and MIRVed SLBMs	1200
(c) Both (a) and (b) plus heavy bombers equipped for long-range cruise missiles	1320
(d) (a)–(c) plus launchers of unMIRVed ICBMs and unMIRVed SLBMs, and heavy bombers not equipped for long-range cruise missiles: aggregate limit to apply from 1 January 1981 and to be achieved by 31 December 1981	2250
(e) Aggregate limit for (d) to apply up to 31 December 1980	2400

(ALCMs) per heavy bomber was agreed, while existing heavy bombers could carry no more than 20 ALCMs. Ceilings on the throw-weights of heavy and light ICBMs were established and various other restrictions were laid down in respect of launchers. (The throw-weight of a missile is the combined weight of the re-entry vehicle(s), the 'bus' used to release or target the re-entry vehicles, and any anti-ballistic missile penetration aids including their release devices.)

Table 2.4 shows the number of strategic arms of the superpowers at the time the Treaty came into force. To comply with the aggregate number of 2,250 the USA had to dismantle 33 vehicles against the Soviet Union's 254. The USA had 1,046 MIRVed

Table 2.4 Strategic offensive arms, June 1979

	USSR	USA
ICBM launchers	1398	1054
SLBM launchers	950	656
Heavy bombers	156	573
Total	2504	2283

launchers (496 Poseidon C-3 SLBMs and 550 Minuteman IIIs) and could therefore deploy an additional 154 MIRVed launchers. The Soviet Union had 608 MIRVed ICBMs and 144 MIRVed SLBMs, and could therefore deploy 448 additional MIRVed ballistic missiles. However, the limitation placed on the deployment of additional RVs on existing ballistic missiles tended to disadvantage the USSR, whose SS-17s, SS-19s and SS-18s had 4, 6 and 10 RVs respectively. By freezing the number of 'heavy' missiles (which were defined as ones with a greater launch-weight or throw-weight than the existing Soviet SS-19 ICBM) the Treaty restricted the Soviet Union from deploying any more heavy ICBMs. (The launch-weight of a missile is the weight of the fully loaded missile at the time of launching.) Since the USA had no heavy ICBMs, this favoured them. Finally, the Treaty allowed the USA to go ahead with the flight-testing and deployment of its new MX but hampered existing Soviet programmes which included a single-warhead replacement for the unMIRVed SS-11 and a new MIRVed ICBM to replace the SS-17 and SS-19.

In spite of these advantages for the USA, the Treaty was heavily criticized by US conservatives who argued that existing silo-based ICBMs were becoming vulnerable to a Soviet first-strike while the Protocol barred the US from adopting a mobile-basing mode to overcome this. Given the US SLBMs and strategic bomber force, the implicit charge that the US nuclear forces were vulnerable to a Soviet first-strike was unfounded. Moreover, in important respects (for example, the deployment of ground- and sea-launched cruise missiles) the Treaty did not constrain the USA, which had programmes in both these fields, beyond 1981. The opposition to the Treaty was such, however, that it was already running into problems at the Senate Committee on Armed Services when the Soviet invasion of Afghanistan (December 1979) led to President Carter's decision to withdraw it from the Senate calendar. In spite of this, both countries acted as if they were governed by it.

Modernization of US Strategic Forces

US strategic forces were in 1984 based on a strategic triad of bombers, ICBMs and SLBMs. There are still 272 B-52 bombers in service: 270 have recently been given a new role as cruise-missile carriers. A replacement bomber, the B-1 will begin to come into service in 1986, and will eventually be converted to carry air-launched cruise missiles (ALCMs) and short-range attack missiles. The ALCM is itself being improved with the planned introduction of a 3,000-mile-range advanced ALCM which will make use of 'stealth technology' – the generic term for various technologies designed to reduce the chance of detection by radar. A stealth-technology advanced technology bomber is also planned for the 1990s. So far as ICBMs are concerned, the backbone of the US force is the Minuteman II (450 missiles) and Minuteman III (550). The Titan II, of which there were 39 in 1984, is being withdrawn.

Planning for a mobile ICBM (the MX) dates from 1966–7; development began in 1974. Construction of 100 MX Peacekeepers has now been approved. A wide-variety of mobile basing-modes have been considered but none as yet has been approved and it is now proposed to base the missiles in existing silos. The real importance of the MX lies in its accuracy, which is likely to be as low as 100 and even possibly 50 yards. This makes it an ideal first-strike war-fighting weapon. Development of a small single-warhead mobile ICBM (dubbed Midgetman) is also under consideration.

The backbone of the US SLBM fleet remains the Poseidon C-3 which was first deployed in 1971 and which is carried on 19 boats, 16 missiles to each boat. Proposals for a replacement SLBM which could be fitted to boats operating from US rather than forward-based ports date from 1966–7; full-scale development of a replacement (the Trident system) was authorized in 1971.

The Trident system comprises a new SLBM and a submarine. The Trident Mark I or C-4 SLBM, with from 7 to 10 MIRVed warheads, is the same size as the Poseidon SLBM but has a range of 4,600 miles (1,700 more than the C-3), thus increasing tenfold the area of sea from which it can hit the USSR. Twenty new Ohio-class boats are being built to carry the Trident C-4, the first of which entered service in 1981. The C-4 is also being retro-fitted to 12 of the Poseidon boats. Advanced development work on another Trident SLBM, the Trident Mark II or D-5, began in 1981. This missile, which is due to be deployed in 1989, will have a range of 6,500 miles and will probably have 7 or 8 warheads.

The most significant development during the 1970s and 1980s has been the improvement in missile accuracy. Accuracy is measured in terms of a missile's circular error probable (CEP), which is the radius of the circle drawn around a target within which half of the warheads aimed at a target are expected to fall. The development of stellar inertial guidance (SIG) systems has enabled on-board navigational computers to correct a missile's course on the basis of a star fix on its position. The missile bus also makes course corrections as it manoeuvres to drop each warhead off for its target, thus increasing accuracy. SIG systems have proved particularly important for SLBMs where the submarine is handicapped by not knowing its exact location at the time of launch.

Research dating from 1964 explored the principle that, if the velocity of a missile in flight and the time needed for a radio signal to travel between it and a satellite are known, then the distance between them can be computed. This led to the Navstar project. The Navstar system will eventually (1987) consist of 18 satellites in six orbital planes providing an accurate location fix to within 53 feet in all three dimensions. When operational this will enable missile accuracies to be brought down to 100 m. Both the Trident II and the MX will be able to use the Navstar, and it may also be retro-fitted into Trident I and possibly even Poseidon missiles.

None of these devices help to improve accuracy by correcting any deviation from the flight path that arises as the warhead re-enters the Earth's atmosphere. This problem has been overcome by developing manoeuverable re-entry vehicles (MARVs) which, in their advanced form, use sensors to scan the target area and correct the warhead's direction. It has also been suggested that individual MARVs may be equipped to receive fixes from Navstar satellites, thus giving an accuracy of about 10 m.

These developments improve the lethality of missiles which is a measure of the ability to destroy hardened military targets, and is a function of (i) the explosive yield of the warhead, (ii) the accuracy of the missile, and (iii) the hardness of the target as measured by its ability to withstand blast pressures. A two-fold increase in accuracy increases the lethality by a factor of four; so does an eight-fold increase in yield. If accuracy can be increased, warhead yield can be diminished.

Soviet Strategic Forces

The Soviet Union still has about 100 ageing Tu-95 Bear turbo-prop bombers fitted with either free-fall bombs or a single Kangeroo air-to-surface turbo-powered cruise missile with a range of 400 miles, and 45 M-4 Bison bombers. Reports of a replacement aircraft have not been substantiated. However, a supersonic penetration/stand-off bomber, the RAM-P Tu-X Blackjack, which is said to have a range of 8,000 miles, an air-launched cruise missile and an air-launched stand-off missile, has been developed.

ICBMs remain the backbone of the Soviet strategic triad. During the 1970s the Soviet Union modernized its ICBM force, introducing the SS-17 (150 missiles), SS-18 (308) and SS-19 (about 300) missiles. Most of these are equipped with MIRVs. Some earlier generation ICBMs are still deployed including the liquid-fuelled SS-11 (some 500 in service) and SS-13 (60). The Soviet Union is testing two new ICBMs. The SS-X-24 is thought by some Western observers to be similar to the SS-19, although Soviet sources say that it is a modification of the SS-13 and hence an allowable development within the spirit of the SALT II Agreement. A smaller ICBM, the SS-X-25, may be mobile. US reports have suggested that a third new missile, the SS-X-26, may be about to be tested, but this has not been confirmed. Although their accuracy has been improved, Soviet ICBMs are generally not as accurate as US ones.

950 Soviet SLBMs are deployed on 62 Hotel, Delta and Yankee class submarines. Soviet submarines have a much poorer record for serviceability than do US ones. Whereas the USA manages to keep over 20 of its 35 SSBNs on patrol at any one time, the USSR can usually manage only about 12 of its 60 boats. The Soviet Union is also deploying its new Typhoon class submarine, the first of which entered service in 1982. The most numerous Soviet SLBM, the SS-N-6, has a range of 2,000 miles and a CEP of about 1,000 yards. The more advanced SS-N-8 (216 missiles) has a range of 5,000 miles. The SS-N-18, with a range of under 5,000 miles is deployed on the Delta class boats (224 missiles) and is also thought to be deployed on the new Typhoon-class submarine.

Theatre Nuclear Forces

By the mid-1960s the Soviet Union had some 630 long-range theatre nuclear forces (LRTNF) directed at Western Europe, including SS-3s, SS-4s and about 880 bombers (the Tu-16 Badger and Tu-22 Blinder). As the missiles became increasingly vulnerable the Soviets experimented with a mobile launcher (the SS-14) and began to deploy SS-11 and SS-19 ICBMs aimed at targets in the European theatre. When development of the SS-16 ICBM was prohibited by SALT it was adapted into a two-stage IRBM, the SS-20, which began to be introduced during 1976–7 in the place of the ageing, inaccurate and vulnerable SS-4s and SS-5s. While the SS-20 armed with three 150-kiloton MIRVed warheads greatly improved Soviet capabilities, its introduction came as no surprise to NATO. By January 1984 some 378 SS-20s had been deployed, one-third of them in the western Soviet Union, 117 east of the Urals, capable of reaching Western Europe and the Far East, and over 100 in Eastern Siberia. Although it is mobile, the SS-20 has to be fired from physically pre-prepared positions and while it is much more accurate than the SS-4s and SS-5s it is replacing, it is still very inferior to the ground-launched cruise missile (GLCM) and Pershing II. The Soviet Union is reported to be developing a new intermediate ballistic missile (SS-X-28), a ground-launched cruise missile (SS-CX-4), and a sea-launched cruise missile. It has begun to deploy a number of submarines off the coast of the USA.

In contrast to the Soviet Union, the USA phased out its vulnerable European theatre nuclear forces (TNF) between 1963 and 1968 and placed its reliance in strategic nuclear systems and Polaris/Poseidon. Fearful that these strategic systems would come to be regarded as weapons of last resort and that the US nuclear guarantee to Europe was being weakened, the West German Chancellor Helmut Schmidt drew attention in 1977 to the existing disparity between NATO and Soviet tactical nuclear and conventional forces. His remarks focused NATO attention on a 'gap' in NATO's deterrent capabilities and in October 1977 NATO's Nuclear Planning Group established a High Level Group to examine the need for NATO TNF modernization. Throughout the official discussions, TNF modernization was seen as 'a function of the deficiencies of flexible response. NATO doctrine requires the capability to strike the Soviet Union with systems based in Europe. Existing capabilities... were aging and were in need of replacement. Modernisation was required irrespective of developments in Soviet capabilities' (North Atlantic Assembly, 1981). Only subsequently was attention focused on the Soviet SS-20s as a means of justifying NATO's deployment of modernized theatre nuclear weapons.

Once the proposal to modernize NATO's TNF gained impetus, it came to be seen in terms of NATO's ability to cope with the challenge of joint-decision making and a test-case of American leadership. In part this was a reaction to the public relations fiasco over the 'neutron bomb' decision which had shown NATO's decision-making to be in disarray. As one observer, Klaas de Vries, commented in a Report to the Military Committee of the North Atlantic Assembly, 'In its extreme form, this view held that whatever military program was eventually decided on would, in the long run, be less important than the taking of the decision itself' (de Vries, 1979). It was against this background that NATO took the decision on 12 December 1979 to deploy 572 US GLCMs and Pershing II missiles in Europe. 464 GLCMs are being deployed. They have a reported range of 1,600 miles. Fitted with a guidance system that combines inertial guidance and terrain contour matching (TERCOM) systems, they will have an accuracy of about 50 m. The Pershing II missile, with a range of about 1,150 or 1,300 miles (depending on whether one believes NATO or Soviet reports) employs terminal guidance systems to provide it with a CEP of about 50 m. It is an ideal first-strike weapon. If its range is as great as 1,300 miles it will be able to reach the Greater Moscow area from its launch sites. Although only 108 missiles are due to be deployed in West Germany, the 1984 US budget allowed for the planned procurement of 311 missiles.

In addition to these weapons, development of sea-launched cruise missiles began in 1972, one of which is a nuclear version with a range of 1,500 miles (the Tomahawk BGM-109A). The USAF also deploys various versions of the F-111 interceptor bomber, including seven squadrons of nuclear capable F-111s based at Upper Heyford and Lakenheath in Britain. Both the US Army and the West German army deploy Pershing IA missiles, the latter operating them under dual key control. Some 350 of these missiles are deployed in all.

The Strategic and Theatre Nuclear Balance

The composition and character of the US and Soviet strategic intercontinental nuclear forces differ widely. Both sides deploy land-based intercontinental ballistic missiles (ICBMs), submarine-launched ballistic missiles (SLBMs) and long-range bombers carrying nuclear bombs and missiles. The Soviet Union has historically placed great emphasis on its ICBM force, which accounts for 65 per cent of its warheads. Submarine-launched ballistic missiles account for another 32 per cent of the warheads and bombers for only 3 per cent. The long-range bomber fleet deployed by the Soviet Union is now 20 to 25 years old. The USSR has fewer aircraft, and few of them are maintained on alert. US policy continues to emphasize the maintenance of a strategic triad of land- , sea- and air-based forces. Some 22 per cent of US warheads are carried by land-based ICBMs, whereas 51 per cent are submarine-based. The US strategic bomber force, like that of the Soviet Union, is now 20 to 25 years old. However, the aircraft have been continually modernized and will be complemented by the deployment of long-range air-launched cruise missiles.

One means of measuring the strategic balance is by counting the number of launchers each superpower has. While this method is used by those who wish to reflect a Soviet advantage (see table 2.5), it fails to take account of the fact that since the 1960s the US has preferred to increase its strategic nuclear capability by capitalizing on its lead in the technology of multiple warheads. A further factor is that the quality of US launchers – ICBMs, SLBMs and long-range bombers – is much better than that of the Soviet Union. This is reflected in the length of time it takes to prepare missiles, particularly the more vulnerable land-based missiles, for firing; the proportion of SLBM submarines at sea; and the readiness of the long-range bomber force. The qualitative lead of US missiles is also reflected in their greater accuracy. This explains the Soviet lead in equivalent megatonnage (EMT) which is another measure frequently used by those who wish to suggest that the Soviet Union is in the lead. As Kaplan (1980, p. 30) has pointed out, stress on the Soviet lead in EMT is 'dishonest because in the mid-1960s the Defense Department made a conscious and deliberate choice to trade in throw-weight for accuracy'. Far from reflecting a Soviet lead, their bigger missiles reflect the fact that Soviet technologists have yet to perfect the 'chip' technology needed for higher performance inertial and terminal guidance systems.

If one measures the total number of strategic warheads available to each superpower one finds that the USA has a significant lead over the Soviet Union (table 2.5). Expressing the strategic balance in terms of warheads reflects the overall capabilities of the

Table 2.5 The strategic balance, May 1984

	Land-based (ICBM)		Sea-based (SLBM)		Gravity bombs SRAM & ALCM		Total	
	US	USSR	US	USSR	US	USSR	US	USSR
Launchers	1,136	1,378	592	949	—	—	—	—
Total strategic warheads	2,236	5,488	5,344	2,309	3,146	290	10,726	8,087

superpowers to destroy the other's military, economic, industrial and command and control structures. A British research group, Scientists Against Nuclear Arms has recently compared the strategic balance in terms of total lethality (SANA, 1984). The results (given in table 2.6) show a significant US lead in missile accuracy. (Accuracy and hence lethality, is not definable in the case of bombers, since it depends on crew performance.)

There is even greater dispute about the theatre nuclear balance, not least because there is no standard accepted definition of what constitutes a theatre nuclear weapon. Kelly (1984, 38–42) presents five views of the theatre nuclear balance. He points out that US figures:

(1) have excluded the Pershing IA from NATO calculations, while including the Soviet SS-12, a weapon comparable to the Pershing IA;
(2) count all Soviet medium-range nuclear-capable aircraft, even though most sources indicate that only about a quarter of them have a nuclear role, while only counting NATO nuclear-capable aircraft that actually fulfil a nuclear role;
(3) exclude British and French forces and Poseidon warheads allocated to NATO;
(4) make no allowance for SS-20s deployed against China.

US figures thus claim a significant imbalance in favour of the Soviet Union whereas the latter suggests that there is an approximate balance between the two superpowers.

Table 2.6 Soviet and US strategic nuclear weapons, 1983

	USSR			USA		
	Number of missiles/ aircraft	Number of warheads	Total lethality	Number of missiles/ aircraft	Number of warheads	Total lethality
Intercontinental ballistic missiles	1,400	4,900	44,000 to 103,000[1]	1,050	2,150	56,300
Submarine-launched ballistic missiles	940	2,800	8,500	640	4,960	14,060
Cruise missiles	—	—	—	190	190	22,600 to 90,500[2]
All missiles	2,340	7,700	53,000 to 112,000[1]	1,880	7,300	93,000 to 161,000[2]
Bomber aircraft	145	290	—	316	2,380	—
Total numbers	2,485	8,000	53,000 to 112,000[1]	2,200	9,700	93,000 to 161,000[2]

Notes: [1] Lower figure assumes accuracy (CEP) of 500 m for SS–19; higher assumes 300 m.
[2] Lower figure assumes accuracy (CEP) of 100 m for the ALCM; higher assumes 50 m.

Source: SANA, (1984). Reproduced with permission

British figures tend to concentrate on land-based systems and hence exclude the NATO and British Poseidon and Polaris systems, and French forces. The US Poseidon missiles illustrate the problems that arise in categorizing weapons. The five Poseidon submarines armed with C-3 missiles that are based at Holy Loch in Scotland are assigned to the Supreme Allied Commander for Europe (SACEUR). They have a range of 4,600 km and have an acknowledged 'theatre' nuclear role. Another 14 US Poseidon submarines with C-3 missiles are not assigned to SACEUR. All 19 Poseidon submarines are SALT-counted, and hence regarded as strategic arms. However, Soviet SS-20s, which have a range of 5,000 kilometres, are regarded as theatre nuclear weapons even though they have a longer range than the Poseidon C-3. The distinction between theatre and strategic weapons is generally meaningless in Europe, and also the Soviet Union. It does, however, have considerable significance for the USA, which cannot be targeted by 'theatre' nuclear weapons deployed in the USSR and other Warsaw Pact countries.

If *all* systems are counted, and if the unit of measure used is the warhead rather than launchers, then a significant NATO advantage can be computed. Figures prepared by SANA (1984) give a Warsaw Pact lead in delivery vehicle/launchers and warheads, but a significant NATO lead in lethality (see table 2.7).

Failure of the Intermediate and Strategic Arms Talks, 1981–3

The NATO decision to modernize theatre nuclear forces was accompanied by an offer to the Soviet Union to establish permanent limitations on European theatre nuclear weapons through arms control negotiations. The Geneva Talks on Intermediate Nuclear Forces (INF Talks) opened on 30 November 1981. In the ensuing months various proposals were put forward, beginning with a NATO offer not to deploy GLCMs and Pershing II missiles if the Soviet Union withdrew all its SS-4s, SS-5s and SS-20s (the 'zero option'). The proposal was described by Soviet Defence Minister Ustinov as absurd. Counter-proposals from the Soviet Union to establish upper limits on TNF foundered because the British and French refused to include their independent nuclear forces in the total, whereas the Soviet Union demanded that they should be taken into account. The deployment of the first GLCMs

Table 2.7 NATO and Warsaw Pact 'theatre' nuclear weapons in Europe, 1983

	Warsaw Pact			NATO		
	Number of missiles/ aircraft	Number of warheads	Total lethality	Number of missiles/ aircraft	Number of warheads	Total lethality
Missiles	440 (620)[1]	920 (1,400)[1]	1,350 (4,900)[1]	351 (431)[2]	479 (1,279)[2]	16,960 (18,710)[2]
Medium-range bombers and strike aircraft	2,700	2,800	—	1,750	2,840	—
Cruise missiles	—	—	—	32	32	15,000
Total numbers	3,140 (3,320)[1]	3,720 (4,200)[1]	1,350 (4,900)[1]	2,130 (2,210)[2]	2,990 (3,790)[2]	32,000 (34,000)[2]

Notes: [1] Larger figure includes 60 SS–19s (360 warheads) and 120 SS–11s (120 warheads); smaller figure does not. Both figures assume ⅔ of SS–4, SS–5 and SS–20 missiles are assigned to the European theatre.

[2] Larger figure includes 5 Poseidon submarines (800 warheads) assigned to SACEUR; smaller figures does not. If only 400 of the Poseidon warheads are counted, the upper figures in the last three columns are reduced by 40, 400 and 1,850 respectively. Note that the contribution of only 9 Perishing II warheads to the lethality is over 13,000.

Source: SANA, (1984). Reproduced with permission

in England (14 November 1983) and Pershing IIs in West Germany (22 November) brought the talks to an end (23 November).

As far as strategic arms were concerned, the Soviet invasion of Afghanistan brought the SALT talks to an end. Both sides nevertheless agreed to adhere to the unratified SALT II Treaty. Conscious of adverse criticism and anxious to defuse the growing American 'freeze' movement, President Reagan announced on 9 May 1982 that the USA would propose a 'practical phased reduction plan' for strategic arms. As a result the Strategic Arms Reduction Talks (START) between the Soviet Union and the USA were convened, each side putting forward different proposals with the USA proposing reductions in warheads and the Soviet Union proposing reductions in delivery vehicles (table 2.8). Progress was so slow that *Newsweek* (10 January 1983) characterized them as 'little more than a polemical exercise' and, in spite of further offers, including a 'build-down' offer in which it was proposed by the USA that two old warheads should be destroyed for each new one deployed, nothing concrete had emerged when, following NATO's deployment of its modernized TNF, the Soviet Union suspended the winter session of the talks on 8 December 1983.

Resumption of Arms Talks, 1985

Following the collapse of the Intermediate and Strategic Arms Talks in 1983 there was considerable speculation throughout 1984 on whether or not they could continue. President Reagan's re-election, by reconfirming his authority, cleared the way for progress towards a resumption of talks and on 6 and 7 January 1985 US Secretary of State George Shultz met Soviet Foreign Minister Andrei Gromyko at Geneva. At this meeting it was agreed that there would be new arms talks with each side fielding a single delegation to conduct three sets of negotiations: on long-range nuclear weapons, intermediate-range nuclear weapons, and weapons in space. Subsequently it was announced that the talks would begin on 12 March. It remains to be seen whether the talks will accomplish anything but the expectation is that they will be a long and arduous process. President Reagan, asked how optimistic he was about an eventual agreement, responded: 'It's hard to be optimistic when you look back at the record' but, on the other hand, 'you'll get an agreement when it is to their [i.e. the Soviets']

Table 2.8 US and Soviet START proposals

	US forces			Soviet forces		
	Existing level	Proposed level	Change	Existing level	Proposed level	Change
US proposals (phase 1)						
Total warheads	7128	5000	−2128	6735	5000	−1735
ICBM & SLBM launchers	1564	850	− 714	2415	850	−1565
ICBM warheads	2152	2500	+ 348	5302	2500	−2802
Soviet proposals						
All delivery vehicles	1940	1800	− 140	2650	1800	− 850

practical interest also. They know that they cannot match us industrially ... And this leads me to believe that possibly they can see the practicality of this [i.e. an agreement] and do it' (Reagan, 1985).

The Impact of Technology on Arms Control

The chances of halting the arms race and reaching an agreement on arms control look increasingly bleak. Recent developments in technology which threaten the other side's retaliatory nuclear forces are inherently destabilizing because they appear to favour the side that strikes first. This undermines the mutual balance of terror, encourages dangerous tendencies to rely on automated response and launch-on-warning, and increases the chance of accidental nuclear war.

The Militarization of Space

The militarization of space began in 1957 with the launch of Sputnik I. Since then a large number of satellites of varying types (reconnaissance, surveillance, early warning, navigational, weather, geodetic, and communications) have been placed in orbit. The military importance of satellites has led to two developments: satellite protection and anti-satellite warfare (ASAT). Satellite protection includes measures to protect ground-based satellite command control and communication systems, data receiving systems, and satellite tracking and surveillance stations; and the hardening of satellites to make them less vulnerable to attack. The latter includes the development of systems that warn satellites of the approach of an unfriendly satellite and enables them to manoeuvre or migrate to higher orbits to escape, and of so-called 'dark satellites' that are less easy to detect.

US interest in ASAT warfare dates from 1959 when the USA successfully launched a missile from a B-47 aircraft and destroyed the Explorer 6 satellite. The Nike Zeus ABM missile was also modified for ASAT purposes and deployed on Kwajalein Atoll in the Pacific between May 1963 and 1967 when the system was decommissioned. Thor IRBMs were modified and deployed on Johnston Island in the Pacific between 1967 and 1973. This system remained nominally operational until 1975. Both systems depended on the explosion of a nuclear warhead to destroy the target satellite. Their range made them of limited effectiveness. It was not until the mid-1970s that further work on ASAT systems began in the USA. Both the Soviet Union and the USA now have extensive research programmes in this field. Current US programmes include launching missiles from high-flying F-15s (a programme which has reportedly not been very successful and which the House of Representatives voted to ban in May 1984), and manoeuvering anti-satellite satellites close to a target one and then exploding it. The Soviet Union has also tested satellite interceptors since 1967, with some 20 tests being conducted in all. More advanced technology, including high-energy beam weapons and high-energy lasers, are under research and development. A number of problems need to be solved before such systems can be used in practical weapons.

Ballistic Missile Defence (BMD)

Under the 1972 Anti-Ballistic Missile Treaty, both the USA and the Soviet Union were allowed two ABM sites, one defending their capital city and the other an ICBM field. In 1974 a Protocol to the Treaty eliminated one of these. The USA currently has three basic networks to warn of a ballistic missile attack: the ballistic missile early warning system (BMEWS) which has recently been improved and which would track ICBMs coming over the Arctic; a series of phased-array radars that would provide warning of an SLBM attack; and early warning satellites capable of tracking rockets during the powered section of their flight. More sophisticated systems are under development.

The US Defense Advanced Research Project Agency (DARPA) has for some time been encouraging research and development of high-energy laser and sub-atomic particle beam-weapons. On 23 March 1983 President Reagan proposed the development of a space-based anti-ballistic missile system. The President called on scientists 'to give us the means of rendering... nuclear weapons impotent and obsolete' by researching into futuristic ballistic missile defence systems. Officially the Strategic Defense Initiative (SDI), the idea was popularly referred to as the 'Star Wars' proposal. Moscow denounced the proposal and attacked Reagan for wanting 'to perpetuate the arms race and carry it over into the 21st century' (*Newsweek*, 4 April 1983). The President, however, saw ballistic missile defence as another way of closing the 'window of vulnerability' which faced US ICBMs and other strategic systems. While he and those supporting his proposal were able to present it as essentially defence-orientated, critics argue that the system would be prohibitively expensive, vulnerable to attack, unlikely ever to be 100 per cent effective, and an abrogation of the 1972 ABM Treaty which was itself premised on the notion of mutual assured destruction (MAD) and the balance of terror.

In the months following President Reagan's statement, two teams of experts, the Defensive Technologies Study Team and the Future Security Strategy Study, prepared separate reports which were then brought together by the Pentagon and the National Security Council to form the basis for a major research and development programme into space-based anti-ballistic missile systems. Following receipt of the report, President Reagan agreed in principle to a space-based anti-ballistic missile system. US$1.4 billion was allocated for R & D into 'star wars' defences in 1985.

The proposals that are now emerging provide for a 'layered' defence involving, firstly the early identification of a missile launch through early warning satellites, ground-based early warning radar, and other sensors, and the use of high-energy laser beams to 'thin out' Soviet ICBMs as they rise from their silos. In the second layer, space-based and land-based lasers operating through large Earth-orbiting mirrors to be launched on warning of an attack would deflect the beams onto incoming warheads. A sounding rocket equipped with optical sensors to locate and track the warheads would meanwhile be launched. This would transmit data on the warheads' trajectories to interceptor rockets, which would rise to attack the incoming warheads in mid-flight. Another proposal calls for the siting of 432 satellites, each armed with 40 M45 interceptor missiles. Development work on non-nuclear kill (NNK) warhead technology and various target direction, recognition and tracking systems is also underway. 'Kills' would be achieved by either laying down a screen of pellets in the path of an incoming warhead which would destroy it on impact, or by intercepting it. In June 1984, following three previous failures, a US interceptor missile launched from Meek Island in the Pacific successfully used infra-red sensors and on-board guidance systems to home in on and then destroy the dummy warhead of a missile that had been launched from California. Finally, in the third layer of defence, point defence systems would locate and try to destroy surviving warheads at low altitude as they closed in on their targets. The US is also working on space-based technologies that would make it possible to defend its NATO allies against shorter-range Soviet missiles (*The Guardian*, 5 February 1984).

US intelligence sources have claimed that the Soviet Union is ahead of the USA in research into ground-based, high-energy lasers for ballistic missile defence, and may be able to test such systems by the early 1990s with deployment following after 2000 or possibly earlier (*The Guardian*, 3 April 1985). The Soviet Union has denied this, stating that the allegation is another example of 'threat inflation' by the US intelligence services (*ibid.*).

There is general agreement that successful development of space weapons technology would radically alter the strategic relationship between the superpowers. Firstly, it undermines the concept of deterrence based on the threat of mutual assured destruction, upon which peace is said to have been based for the past 40 years. The most serious objection to the scheme is that no one appears to believe that a fool-proof system can be devised. More importantly, the chances of successful arms-control negotiations will be jeopardized since deployment of BMD systems, although not necessarily the research into them, will contravene the spirit if not the letter of earlier arms-control treaties, upset the existing balance of power, and will make a surprise attack apparently more feasible and therefore, at least potentially, more likely.

Command, Control, Communication and Intelligence (C^3I)

Considerable attention has been paid by the USA to the development of reliable command, control, communication and intelligence (C^3I) systems which will ensure the continuation of military command and control during a nuclear war. The Worldwide Military Command and Control System (WWMCCS, pronounced 'Wimex') consists of a network of underground and airborne command posts including duplicate command and control systems and a communication system protected against the atmospheric effects of nuclear explosions and intentional jamming. Satellites play a key role in the provision of information through reconnaissance and early warning and in communication.

The early Wimex network had serious shortcomings. During 1979–80 President Carter's administration initiated a wide range of measures to improve C^3 systems. Presidential Directive PD-53 'National Telecommunications Policy', signed by Carter on 15 November 1979, proclaimed that 'it is essential to the security of the US to have telecommunications facilities adequate to satisfy the needs of the nation during and after a national emergency ... to provide continuity of essential functions of government, and to reconstitute the political, economic and social structure of the nation.' The directive requires that the communications system 'support flexible execution of retaliatory strikes during and after an enemy attack'. PD-58, signed by Carter on 30 June 1980, directed the Department of Defense and other government agencies to improve the capacity of certain parts of the government machine to withstand a nuclear attack. PD-59, signed by Carter on 25 July 1980, recognized that the current US C^3 system was inadequate to support a policy of extended nuclear war-fighting and noted that the strategic policy outlined in the directive required improvements in C^3I 'in areas of increased flexibility and higher assurance of command-and-control survivability and long-term endurance' (Senate Committee on Armed Services, 1981).

While PD-53 and PD-58 paved the way, it is the Reagan Administration that has taken steps to improve C^3I by improving early warning systems, providing more survivable command posts, and improving communication systems. The aim is that these should be 'capable of supporting controlled nuclear counter-attacks over a protracted period while maintaining a reserve of nuclear forces sufficient for trans- and post-attack protection and coercion' (*Washington Post*, 10 November 1982). The improvements the Reagan administration intends to make are wide-ranging and cover all the components of the C^3I system. Indeed, two C^3I systems are being developed, a peacetime/crisis/conventional war system which, while it is being hardened to increase its probability of survival, is not expected to remain intact during a nuclear war, and a nuclear war C^3I system which is designed to survive and endure such a war (Richelson, 1983).

Anti-Submarine Warfare (ASW)

The objective of strategic ASW is to destroy the enemy's submarines before they can launch their missiles. The USA has a commanding lead in this field using a variety of sensors to detect and track enemy boats. From an arms control point of view, such developments are highly destabilizing since they endanger the Soviet Union's retaliatory capability.

References and Further Reading

Acheson, D. (1970) *Present at the Creation*, New York: New American Library.

Bush, V. (1970) *Pieces of the Action*, New York: William Morrow.

de Vries, K. (1979) 'The Role of Nuclear Weapons in Alliance Strategy', Report to the Military Committee of the North Atlantic Assembly, November 1979. Cited in US Congress Research Service (1981) 'The Evolution of NATO's Decision to "Modernize" Theatre Nuclear Weapons', in Alva Myrdal et al. (1981) *Dynamics of European Nuclear Disarmament*, Nottingham: Spokesman, 100–01.

International arms control treaties and agreements:

US-Soviet Treaty on the limitation of anti-ballistic missile systems, 1972 (SALT ABM Treaty).

US–Soviet Interim Agreement on certain measures with respect to the limitation of strategic offensive arms, 1972 (SALT I Agreement).

Protocol to the US–Soviet Treaty on the limitation of anti-ballistic missile systems, 1974.

Joint US–Soviet Statement on the question of further limitations of strategic offensive arms, 1974 (Vladisvostok Accord).

Treaty between the USA and the USSR on the limitation of strategic arms, 1979 (SALT II Treaty).

Protocol to the SALT II Treaty, 1979.

Kaldor, M. (1981) 'Misreading ourselves and others', in K. Coates (ed.) (1981) *Eleventh Hour for Europe*, Nottingham: Spokesman.

Kaplan, F. M. (1980) *Dubious Specter: A Skeptical Look at the Soviet Nuclear Threat*, Washington: Institute for Policy Studies.

Kelly, A. (1984) *Not by Numbers Alone. Assessing the Military Balance*. London: Housmans/Bradford: University of Bradford School of Peace Studies.

North Atlantic Assembly (1981), 'Interim Report on Nuclear Weapons in Europe', Report submitted to the US Senate Foreign Relations Committee, December 1981.

NSC-68 (1950) 'United States Objectives and Programs for National Security' (14 April), Department of State, Foreign Relations of the United States, 1950, Vol. 1: *National Security Affairs; Foreign Econo-*

mic Policy, Washington, DC: US Government Printing Office, 1977, 281–2.
Presidential Directives, classified documents:
PD-53 (1979) *National Telecommunications Policy*.
PD-58 (1980) *Continuity of Government*.
PD-59 (1980) *Nuclear Weapons Employment Policy*.
Reagan, R. (1985) cited in *Newsweek*, 18 March.
Richelson, Jeffrey (1983) 'PD-59, NNDD-13 and the Reagan Strategic Modernization Program', *Journal of Strategic Studies*, 6 (2), 125–46.
Rogers, P. (1984) *Guide to Nuclear Weapons, 1984–85*, Bradford: University of Bradford School of Peace Studies.
Rosenberg, D. A. (1983) 'The Origins of Overkill. Nuclear Weapons and American Strategy, 1945–1960', *International Security*, 7 (4), 3–71.
SANA (Scientists Against Nuclear Arms) (1984) *The Nuclear Balance*, Milton Keynes: SANA.
Security Resources Panel of the Science Advisory Committee (1957) *Deterrence and Survival in the Nuclear Age*, 7 November.
Senate Committee on Armed Services (1981) *Department of Defense Authorization for Appropriations for Fiscal Year 1982*. Washington DC: US Government Printing Office.
Stimson, H. L. (1945) Diary entry for 9 August 1945. Cited in Herken, G.F. (1974) *American Diplomacy and the Atomic Bomb*, unpublished doctoral dissertation, Princeton University, 68.
Stimson, H. L. and Bundy, McG. (1947) *On Active Service in Peace and War*, New York: Harper.
US Congress Research Service (1981) 'The Evolution of NATO's Decision to "Modernize" Theatre Nuclear Weapons', in Alva Myrdal et al. (1981) *Dynamics of European Nuclear Disarmament*. Nottingham: Spokesman.
Yergin, D. (1980) *Shattered Peace. The Origins of the Cold War and the National Security State*, Harmondsworth: Penguin.

3
US and Soviet Nuclear Policy

Basic Concepts

Nuclear weapons may be used in a variety of ways. The main distinctions are firstly between *first-strike* (a disarming attack on the enemy) and *second-strike* (where the weapons are used in retaliation following a nuclear attack by another state); and secondly between *counter-force targeting* (where the weapons are aimed at the military forces of the other side, particularly at strategic or theatre nuclear forces) and *counter-value targeting* (where they are aimed at 'soft' urban, industrial and economic targets). The latter is also referred to as '*counter-city*' or '*counter-people*' targeting. Counter-force targeting needs very accurate missiles to be successful against hardened missile silos and command and control posts. Where both sides have the ability to absorb a first-strike and deliver a retaliatory second-strike capable of inflicting an unacceptable level of damage, a situation of *mutual assured destruction* is said to exist. The retaliatory attack is normally held to be aimed at counter-value targets since this is likely to do greatest damage to the other country's society, and hence have the greatest deterrent effect, and there is no point in targeting empty missile silos and airfields.

The concept of deterrence is usually understood to mean nuclear deterrence. However, in NATO doctrine there is an assumption that deterrence is invoked across the whole spectrum of military forces, from conventional forces through battlefield and theatre nuclear weapons to strategic ones. This *applied deterrence* requires a wide range of weapons to prepare for different eventualities. It is linked with the strategic doctrine of *flexible response*, requires an ability to destroy military as well as urban-industrial targets, and differs fundamentally from the concept of *pure deterrence* embodied in the idea of massive retaliation.

US Nuclear Policy

Under Truman

In the immediate post-war years the USA enjoyed a monopoly of nuclear weapons. US Air Force planners, drawing on their experience of strategic bombing, believed that air power could be most effectively used to attack the enemy's war-supporting industrial base and communications system and undermine civilian morale, thus leading to the breakdown of the social structure and the collapse of the will to resist. By 1949 this policy was being challenged. The morality of strategic or area bombing had already been questioned on the grounds that non-combatants were its inevitable victims. The development of nuclear weapons increased anxiety on this score. More importantly, the political effectiveness of the strategy was queried. An *ad hoc* committee chaired by Air Force Lieutenant General H.R. Harmon reported to the Joint Chiefs of Staff in May 1949 that the effect of War Plan Trojan would not in itself 'bring about capitulation, destroy the roots of Communism, or critically weaken the power of Soviet leadership to dominate the people' (Rosenberg, 1983, p. 16). Nevertheless, since the atomic bomb was 'the only means of inflicting shock and serious damage to vital efforts of the Soviet war-making capacity', Harmon concluded that 'the advantages of its early use would be transcending'.

These ideas, coupled with the Soviet explosion of an A-bomb, led American planners to expand and divide the target list. Top priority was given to 'the destruction of known targets affecting the Soviet capability to deliver atomic bombs'. Secondary priority was given to 'retardation' targets – that is transportation links whose destruction would slow the westward movement of Soviet forces. Third priority was assigned to attacks on industrial facilities, notably Soviet liquid fuel, electric power and atomic energy industries. However, the majority of targets incorporated into the November 1950 target list continued to fall in the third category. Surprisingly, little urgency was given to identifying counter-force targets. General Curtis LeMay, commander of SAC, felt that isolated target complexes were difficult for aircrew to locate. He argued that 'we should concentrate on industry itself which is located in urban areas' so that, even if the target were missed, 'a bonus will be derived from the use of the bomb' (LeMay, *Diary*, 23 January 1951, cited in Rosenberg, 1983, p. 18).

Conditions under which the USA might use atomic weapons against the Soviet Union were not laid down. National Security Council document NSC-30 (16 September 1948) indicated only that the military 'must be prepared to utilize promptly and effectively all appropriate means available, including atomic weapons'. The NSC noted in April 1950 that 'the military advantages of landing the first blow ... require us to be on the alert in order to strike with our full weight as soon as we are attacked, and if possible, before the Soviet blow is actually delivered' (NSC document NSC-68, pp. 281–2).

Under Eisenhower

Declaratory policy: massive retaliation

President Eisenhower entered office with a greater knowledge of nuclear weapons than any American president since the Second World War. He was an early advocate of nuclear war planning, a firm supporter of increased nuclear production and the man who, as first NATO Supreme Allied Commander Europe, had encouraged planning for the tactical nuclear defence of Europe.

By the end of October 1953 his Administration was in a position to approve a new national security policy (National Security Council paper NSC-162/2) under which America would respond to large-scale aggression with massive nuclear retaliation. Eisenhower publicly enunciated this policy in his State of the Union speech on 7 January 1954. A few days later, Secretary of State John Foster Dulles outlined the policy in greater detail in a speech before the Council on Foreign Relations (12 January 1954). Massive retaliation, said Dulles, involved the 'great capacity to retaliate, instantly by means and at places of our choosing' (1954a, p. 108). He felt that the policy provided 'maximum deterrence at a bearable cost' by placing 'more reliance on deterrent power and less dependence on local defensive power'. Since the time and place of retaliation would remain unclear Dulles felt that the strategic initiative would continue to rest with the USA. Even so, he recognized that 'massive atomic and thermonuclear reaction is not the kind of power which could most usefully be evoked under all circumstances', and he repudiated the notion that the US 'intended to rely wholly on large-scale strategic bombing as the sole means to deter and counter aggression' (Dulles, 1954b, cited in Freedman, 1981, p. 86).

The policy of massive retaliation came to be questioned once it became clear that the Soviet Union might have the power to launch a pre-emptive attack on the USA. In May 1954 a paper prepared by the Joint Chiefs of Staff Advance Study Group proposed that the US consider 'deliberately precipitating war with the USSR in the near future' before Soviet thermonuclear capability became a 'real menace' (Rosenberg, 1983, p. 34). Dulles, however, stated categorically that 'any idea of preventive war was wholly out of the question as far as the United States was concerned' (*New York Times*, 10 November 1954), and in December Eisenhower approved NSC-5540/1, which said that 'the United States and its allies must reject the concept of preventive war or acts intended to provoke war.'

If the idea of preventive war was ruled out, the idea of a pre-emptive first-strike was not. The CIA believed that it would have several days' if not weeks' warning of Soviet preparations for a first strike against the USA. It was assumed that during this period the USA would be able to launch a pre-emptive strike. Moreover, even if the Soviet Union struck first, there would still be time to blunt the Soviet's ability to launch further attacks since it was estimated that it would take up to thirty days for the Soviet Union to deliver all its nuclear weapons (Rosenberg, 1983, p. 34).

Targeting policy

The debate on targeting policy initiated in 1949 continued throughout the Eisenhower period. Air Force chiefs remained 'decidedly hostile to any notions that it might be necessary to find ways of limiting war by eliminating strategic bombing' (Brodie, 1973, p. 394). They nevertheless made use of the plans to target Soviet strategic forces to expand the target list and to justify USAF appropriations for more aircraft. Eventually Army Chief of Staff Maxwell Taylor and Chief of Naval Operations Arleigh Burke prepared a devastating critique of SAC's war plan (August 1957) which indicated that far too many weapons were being assigned to targets. By December that year they were arguing that SAC was vulnerable to a Soviet attack and that in view of this an alternative target list was needed, consistent with the survival of a quarter of the USA's strategic bomber force, which would ensure the 'destruction of government controls and population centres within the USSR to the extent necessary to neutralise the capabilities of the USSR to carry on the war' (Memorandum for General Twining et al, 19 December 1957, cited in Rosenberg, 1983, p. 53). As a result, the concept of a strictly retaliatory nuclear strategy was introduced into Joint Chiefs of Staff guidance for the first time when, in early 1958, the JCS approved the alternative targeting plan as a basis for drawing up target lists.

The increased emphasis on counter-city targeting was strongly supported by Admiral Burke who believed both that the USA could not hope to disarm the Soviet Union in a counter-force attack and that US land-based strategic forces were increasingly vulnerable to a Soviet ICBM first-strike. Burke used these arguments to propose a force of nuclear armed submarines with the 'ability to destroy major urban areas' and thus deter a Soviet attack (Admiral Arleigh Burke, CNO Personal Letter No 5 to Retired Flag Officers, cited in Rosenberg, 1983, p. 186). Air Force chiefs criticized the Navy's strategic proposals. The Air Force Chief of Staff, Thomas D White, argued in March 1959 that an adequate deterrent required a capability both 'to neutralize that military strength which is a threat to the US' whenever warnings of an impending strategic attack allowed time for a response and to respond against 'the Soviet control structure and basic national strength', as well as against any surviving Soviet strategic forces (White, Memorandum for DCS/Plans and Programs, 30 March 1959, cited in Rosenberg, 1983, p. 58).

By 1959 the need to coordinate strategic planning was becoming obvious. The Chairman of the Joint Chiefs of Staff, Air Force General Nathan Twining, believed that the solution was to establish a clear national targeting policy and then to develop a 'single integrated operational plan' to form the basis for the war plan (Rosenberg, 1983, p. 61). Eisenhower strongly supported this proposal and in August he approved the creation of the Joint Strategic Target Planning Staff (JSTPS) under SAC direction, to prepare a National Strategic Target List and Single Integrated Operational Plan. (The NSTL contains data on all the targets that might need to be attacked. The SIOP assigns targets to all strategic weapon systems.) By December 1960 the first NSTL and SIOP had been produced. 2,600 separate installations were selected for attack out of a target list of 4,100. A total of 1,050 designated ground zeros for nuclear weapons were identified, including military and urban-industrial targets. Given sufficient warning, the USA would launch its entire strategic force carrying 3,500 nuclear weapons against the Soviet Union, China and Soviet satellite states (Rosenberg, 1983, p. 6). In essence, all available forces were to be used to achieve maximum destruction in one fell swoop.

Navy leaders criticized the plan. Admiral Burke complained that 'counterforce receives higher precedence than is warranted for a retaliatory plan, and less precedence than is warranted for an initiative plan' (Message, November 1960, cited in Rosenberg, 1983, pp. 7–8). Eisenhower, learning of the Navy's concern, sent his science adviser George B. Kistiakowsky to visit SAC headquarters. On his return he reported that the 'damage criteria and the directives to the planners are such as to lead to unnecessary and undesirable overkill' (Kistiakowsky, 1976, cited in Rosenberg, 1983, p. 8). Eisenhower told his naval aide that Kistiakowsky's findings 'frighten the devil out of me' (Transcript, Admiral Burke's conversation with Captain Aurand, 25 November 1960, cited in Rosenberg, 1983, p. 8) but, as out-going President, he could only leave the problem to his successor.

Limited nuclear war

Since 1949, the United States has been linked to Western Europe through the NATO military alliance. Although the alliance accepted a conventional force goal of 96 divisions at the meeting of NATO ministers held in Lisbon in February 1952, it quickly became apparent that it was not prepared to meet the costs involved. In December 1954 NATO reduced the force level objective from 96 to 30 active divisions. As a result it came to rely on tactical as well as strategic nuclear weapons to deter a Soviet attack on Europe. Only later did two worries emerge. Firstly, the assumption, adopted as policy by the Alliance, that Soviet aggression in Europe would result in massive retaliation came to be questioned by those who wondered whether the USA would risk nuclear annihilation in the defence of Europe. Secondly, it was realized that use of nuclear weapons in Europe would be enormously destructive, particularly in Germany (see p. 15).

By the beginning of 1956 the US Joint Chiefs were deeply divided over the strategic concepts that underpinned the Administration's policy. This policy, laid down in NSC-5602/1 (mid-March 1956) declared that nuclear weapons 'will be used in general war and in military operations short of general war as authorised by the President' (cited in Rosenberg, 1983, p. 42). On the 30 March Eisenhower told the Joint Chiefs that 'any war in which Russian troops were involved directly against United States forces or the United States,' would be a general war, and any such Soviet attack would be met by launching SAC 'as soon as he found out that Russian troops were on the move' (Rosenberg, 1983, p. 42).

In May that year Army Chief of Staff Maxwell Taylor met Eisenhower. Taylor argued that by 1960, with the growth of thermonuclear stockpiles, 'a situation of mutual deterrence must be envisaged'. The greater danger would then be that an initially small conflict might escalate into a general nuclear war. He urged the President to reconsider the use of nuclear weapons in local conflicts. Eisenhower was unconvinced. He did not believe that the use of tactical nuclear weapons would provoke escalation. Taylor also argued against the tendency to focus defence planning on a full-scale nuclear war at the expense of conventional forces. Again Eisenhower did not agree. The Soviets, Eisenhower stated, would certainly use atomic weapons at once 'and in full force'. The only prudent course would be 'to get our striking force into the air immediately upon notice of hostile action by the Soviets'. Meanwhile, the ability to retaliate massively was the key to survival (Rosenberg, 1983, p. 42). In 1958 Taylor once again raised the policy issue. This time he was supported by the Navy and the Marine Corps at Joint Chiefs of Staff level, and by Secretary of State Dulles when the issue was raised at the National Security Council. Dulles wanted to place more emphasis on conventional forces and the possibility of local war. Eisenhower, however, demurred on grounds of cost.

Civilian strategists had also begun to question the over-reliance on nuclear weapons implicit in Eisenhower's 'New Look' defence policy, and particularly the doctrine of 'massive retaliation'. Adlai Stevenson criticized the policy on the grounds that it presented the USA with 'the grim choice of inaction or a thermonuclear holocaust', and invited Moscow to 'nibble us to death' (cited in Coral Bell, 1957, p. 99) while Henry Kissinger concluded that massive retaliation 'makes for a paralysis of diplomacy' (Kissinger, 1955, p. 425). Strategists such as Bernard Brodie and Robert Osgood, as well as Kissinger, nevertheless felt that tactical nuclear weapons provided a means of retaining the benefits of nuclear weapons without running the risk of all-out nuclear war, and they and other writers proceeded to explore the idea of limited nuclear war.

These ideas were by no means universally accepted. During 1956 a book written by members of the Princeton Center for International Studies and edited by William Kaufmann (1956) argued that the underlying assumption upon which much of American strategic theory was based was a continuing monopoly of US power. If one assumed parity between the USA and the Soviet Union, then much of the utility of nuclear weapons fell by the board. This was true of tactical nuclear weapons as well, since their use would greatly increase the risk of general nuclear war. James King (1957) argued that there would be grave difficulties in keeping a nuclear war limited. By 1959 Bernard Brodie had also changed his mind: 'a people saved by us through our free use of nuclear weapons over their territories would probably be the last that would ever ask us to help them' (Brodie, 1959, cited in Freedman, 1980, p. 114). By 1960 Kissinger was also converted: 'The notion that nuclear weapons can substitute for numerical inferiority has lost a great deal of its validity' (Kissinger, 1960, in Brennan, 1961, p. 145). He now favoured strengthening conventional forces and drawing a clear distinction between nuclear and conventional weapons.

The McNamara Years, 1961–8

Strategic policy

The Kennedy Administration came to power committed to expanding American strategic nuclear forces. The new Defense Secretary, Robert McNamara, initiated a wide-ranging review of US defence policy early in 1961. In January 1962 he made his first public reference to the new 'second strike counterforce' policy (Sloss, 1981, p. 7). In May he outlined it at a secret meeting of NATO ministers in Athens and on 16 June he described it in almost identical terms to an audience at the University of Michigan at Ann Arbor:

> The US has come to the conclusion that to the extent feasible, basic military strategy in a possible general nuclear war should be approached in much the same way that more conventional military operations have been regarded in the past. That is to say, principal military objectives, in the event of a nuclear war stemming from a major attack on the Alliance, should be the destruction of the enemy's military forces, not of the civilian population... We are giving a possible opponent the strongest imaginable incentive to refrain from striking our own cities (Kaufmann, 1964, p. 116).

Having been briefed on the 1962 SIOP in February 1961, McNamara also initiated a review of targeting policy. He was particularly disturbed by the rigidity of the plan, the absence of any clear rationale for the mix of targets, and the level of fall-out and destruction implicit in its execution. The results of the review were embodied in the Pentagon's *Guidelines for Planning* which then became the basis for a new SIOP (SIOP-63) that was approved in June 1962 (Ball, 1983, p. 10). These identified five principal 'options' for attack, together with a range of sub-options. The principal ones were:

1 Soviet strategic nuclear delivery forces including missile sites, bomber forces and submarine bases;
2 other elements of Soviet military forces and resources centred away from cities;
3 Soviet military forces and resources near cities;
4 Soviet command and control centres and systems;
5 urban-industrial centres.

Almost as soon as it was announced this strategy of second-strike counter-force came under heavy criticism. Congressional hawks argued that a 'no-cities' policy was a sign of weakness. Doves argued that it made it more likely that the USA would initiate a first-strike and that a second-strike retaliatory attack against military targets (a counter-force attack) was indistinguishable from a US first-strike strategy. Indeed, it would be more logical to strike first against Soviet military targets if the aim was to limit the damage to US cities, rather than to wait until the missile launch pads and silos, and strategic airfields, were empty. The Doves' concern was given substance by President Kennedy's comment that 'Khruschev must not be certain that, where our vital interests are threatened, the United States will not strike first' (Kennedy, 1962). Indeed throughout the debate, officials did not deny that the USA might use its controlled response capability to initiate a strategic nuclear attack on the Soviet Union.

Soviet spokesmen denied that a nuclear conflict could be controlled and reaffirmed their intention to strike at US military and civilian targets simultaneously as soon as nuclear war began. McNamara himself believed that a Soviet nuclear attack on the USA would always include an attack on major urban areas. The fact that the Soviet Union then had so few missiles compared with the USA also meant that it could not afford to enter into a reciprocal counter-force war with the USA.

The USA's European allies also disliked the 'no-cities' policy because it raised the possibility that the USA would separate European security from its own and, by removing the threat of a US strike against Soviet cities, do away with the very deterrent that kept the Soviet Union from invading Europe. From this point it was a short step to envisage the possibility of a nuclear war fought over European territory with the Soviet Union and USA left unscathed.

By late 1962 it was clear that the USAF, while sceptical about any policy of restraint in strategic planning, was using the no-cities counter-force policy to request both more Minuteman ICBMs and the procurement of a force of supersonic reconnaissance strike bombers (the RS-70). To curb this expansion, McNamara shifted to a new policy of 'assured destruction'. The change in policy took place gradually. In January 1963 William Kaufmann, acting on McNamara's instructions, told Air Force generals that 'they were no longer to take the avowed US strategy as a criteria for strategic force proposals' (Ball, 1983, p. 13). In the same month McNamara noted that 'a very large increase in the number of fully hardened Soviet ICBMs and nuclear-powered ballistic missile launching submarines would considerably detract from our ability to destroy completely the Soviet strategic nuclear forces' (McNamara,

1963a). Meanwhile, Pentagon officials began to acknowledge the difficulty of limiting damage to the USA in the event of a major strategic war between the USA and the USSR. On 18 November McNamara acknowledged that such a war would be 'highly destructive to both sides [under] all foreseeable circumstances' (McNamara, 1963b, p. 8).

During 1965 and 1966 McNamara began to emphasize the 'assured destruction' aspects of US strategic policy. Early in 1965 he stressed that, of the two objectives of damage limitation and assured destruction, the latter was the more important and that 'the destruction of, say, one quarter to one third of [the Soviet Union's] population and about two thirds of its industrial capacity would mean the elimination of the aggressor as a major power for many years' (McNamara, 1965). Once an adequate capability for assured destruction had been secured, additional forces would have to be justified in terms of their contribution to the secondary objective of damage limitation. This, however, implied a costly and 'open-ended' commitment.

By 1967 McNamara had taken his argument a step further when he said:

> Damage limiting programs, no matter how much we spend on them, can never substitute for an Assured Destruction capability in the deterrent role. It is our ability to destroy an attacker as a viable 20th century nation that provides the deterrent, not our ability to partially limit damage to ourselves (McNamara, 1967, pp. 38–9).

Damage limitation was, he went on, a wasteful policy because it would only encourage the arms race in offensive and defensive weapons systems. By 1968, he was proclaiming the inevitability of 'mutual deterrence'.

> For a 'Damage Limiting' posture to contribute to the deterrent... it would have to be extremely effective, i.e. capable of reducing the damage to truly nominal levels... we now have no way to accomplish this (McNamara, 1968, p. 47).

However, in spite of his wish to curb excessive expenditure and check demands for force increases and for the development of follow-on systems, McNamara continued to stress the need for US superiority. Moreover, no substantive revisions were made to

SIOP-63. The majority of designated targets were still military installations and options for avoiding attacks on cities were retained. As Henry Rowen commented in his report to the Commission on the Organization of the Government for the Conduct of Foreign Policy:

> The primary purpose of the Assured Destruction capabilities doctrine was to provide a metric for deciding how much force was enough: it provided a basis for denying service and Government claims for more money for strategic forces... However, it was never proposed by McNamara or his staff that nuclear weapons actually be *used* in this way (Rowen, 1975, p. 227).

Morton Halperin, an Assistant Secretary of Defense in the last years of the Johnson Administration, told Desmond Ball in private correspondence that:

> The SIOP remains essentially unchanged since then [McNamara's Ann Arbor speech of 15 June 1962]. There have been two developments, however: 1) it has become more difficult to execute the pure-counterforce option, and its value is considered to be diminishing and, 2) all public officials have learned to talk in public only about deterrence and city attacks. No war-fighting, no city-sparing. Too many critics can make too much trouble (no-cities talk weakens deterrence, the argument goes), so public officials have run for cover. That included me when I was one of them. But the targeting philosophy, the options and the order of choice remain unchanged from the McNamara speech (Ball, 1983, p. 15).

Theatre nuclear strategy: 'flexible response'

When McNamara came to office NATO was firmly wedded to the use of nuclear weapons to offset the weakness of its conventional forces. However, according to Kaufmann McNamara recognized that even low-yield nuclear weapons were extremely destructive 'and hardly the preferred weapons to defend such heavily populated areas as Europe'. Their use also presented 'a very definite threshold, beyond which we enter a vast unknown', with all the dangers of escalation to global nuclear war (Kaufmann, 1964, p. 97).

McNamara wanted to increase NATO's conventional strength and to place tighter controls on the use of nuclear weapons. He also resisted any further development of tactical nuclear weapons. Writing in 1983 he explained that 'the new policy was the result of a recognition by U.S. civilian and military officials that NATO's vastly superior nuclear capabilities... did not translate into usable military power' (McNamara, 1983, p. 64). The build-up of NATO's conventional forces would, it was hoped, improve the alliance's conventional capabilities to the point at which 'the use of nuclear weapons would be unnecessary...[and]... even if this expectation turned out to be false, any use of nuclear weapons would be "late and limited" (McNamara, 1983, pp. 63–4).

McNamara's strategy, which he put to NATO ministers when they met in Athens in April 1962, envisaged that nuclear weapons would have only two roles in the NATO context:

1 to deter the Soviets from starting a nuclear war;
2 were a war to break out and if conventional forces proved insufficient to hold a Soviet advance, to act as a weapon of last resort to persuade the Soviets to end the conflict on acceptable grounds.

These proposals met with strong opposition, firstly, from those who felt that the proposed build-up of conventional forces was too expensive; secondly, from those who felt that any attempt to limit nuclear warfare and prevent it from escalating to a strategic level was an attempt to 'decouple' the USA from the defence of Europe; and, thirdly, from those who felt it increased the chances of highly destructive conventional wars in Europe. The argument raged for five years and was not resolved until 1967 when NATO adopted the strategy of flexible response which was then set forth in a document entitled *Overall Strategic Concept for the Defense of the NATO Area*, known colloquially as MC 14/3 (NATO, 1968).

The revised strategy proposed to deter aggression by maintaining forces adequate to counter an attack at whatever level the aggressor chose to fight. The basic elements of the policy are conventional and nuclear deterrence, forward defence and flexible response involving the will and ability to escalate as necessary, including the first-use of nuclear weapons, thus forcing the aggressor to confront costs and risks disproportionate to his initial objectives. Emphasis continued to be placed on the early use of tactical nuclear weapons, the expectation being that NATO would not be able to fight a prolonged conventional war and that it would be a matter of days rather than weeks before NATO had to use nuclear weapons. The ultimate threat of escalation to a strategic nuclear exchange between the USA and the Soviet Union was, at least in theory, retained as an option.

Under Nixon and Ford

On 21 January 1969, the day after he entered office, President Nixon's National Security Advisor Henry Kissinger initiated a review of the United States' military position. The first hint of a shift of policy came in the President's 1970 Foreign Policy Message to Congress (17 February 1970) when he mused:

> Should a President, in the event of a nuclear attack, be left with the single option of ordering the mass destruction of enemy civilians, in the face of the certainty that it would be followed by the mass slaughter of Americans? Should the concept of assured destruction be narrowly defined and should it be the only measure of our ability to deter the variety of threats we may face? (Nixon, 1970, p. 122).

The following year's Foreign Policy Message (25 February 1971) indicated the thrust of the policy review:

> I must not be – and my successors must not be – limited to the indiscriminate mass destruction of civilians as the sole possible response to challenges (Nixon, 1971, p. 170).

Strategic planning, Nixon argued, could not be based 'solely on some finite – and theoretical – capacity to inflict casualties presumed to be unacceptable to the other side'.

As a result of a series of studies between 1971 and 1973 the Administration was in a position to agree a new strategic doctrine by early 1974. On 17 January Nixon signed National Security Decision Memorandum NSDM-242. It began:

> I have reached the following decisions on United States policy regarding planning for nuclear weapons employment. These decisions do not constitute a major new departure in US nuclear strategy; rather they are an elaboration of existing policy.
> ... The fundamental mission of US nuclear forces is to deter nuclear war and plans for the employment of US nuclear forces should support this mission (cited in the *Washington Post*, 12 October 1980).

The Memorandum went on to direct that further plans 'for limited employment options' be developed and incorporated into the SIOP. Crucial to the policy was the concept of escalation control. It directed that the USA must have the potential to 'hold some vital enemy targets hostage to subsequent destruction' and to control 'the timing and pace of attack execution, in order to provide the enemy opportunities to consider his actions' (cited in the *Washington Post*, 12 October 1980).

The political guidelines contained in NSDM-242 were embodied in the 'Policy Guidance for the Employment of Nuclear Weapons' and the associated 'Nuclear Weapons Employment Policy' (NUWEP) document which was signed by Secretary of Defence, James Schlesinger on 4 April 1974. The latter, also known as NUWEP-1, provided the framework within which the Joint Chiefs of Staff gave guidance to the Joint Strategic Target Planning Staff for the development of a new SIOP, SIOP-5. This was formally approved in December 1975 and took effect from 1 January 1976. It essentially confirmed the targeting policy of the 1963 SIOP while incorporating a few new features including, firstly, the idea that an objective of targeting policy should be to destroy the 'enemy's post-war power, influence and ability to recover... as a major power' (*Washington Post*, 12 October 1980) while keeping cities as ultimate hostages; and, secondly, the idea of 'withholds' or 'non-targets', notably localities which should not be hit under any conditions and centres of political leadership and control whose survival was necessary for the purpose of intra-war deterrence and bargaining. It did not mean that cities were no longer to be targeted. Indeed, the 200 largest Soviet cities and 80 per cent of the 886 cities with populations over 25,000 are targeted simply because of their associated military and industrial facilities.

The extent to which the Nixon Administration was prepared to articulate its new doctrine publicly was another matter. Apart from Nixon's Foreign Policy Statements, there had only been hints that a change of policy was imminent. On 10 January 1974, however, in a speech given at an Overseas Writers' Association luncheon in Washington, Schlesinger declared that the USA could not 'allow the Soviets unilaterally to obtain a counter-force option which we ourselves lack' (Schlesinger, 1974a, p. 86). In spite of the fact that US war plans already targeted Soviet military and strategic installations, Schlesinger said only that 'Military targets, whether silos or other military targets, are, of course, one of the possible target sets' (ibid.). However on 23 January, at a news conference, he admitted: 'We have regularly targeted military targets' (Schlesinger, 1974b).

On 5 February, asked by the Senate Armed Services Committee whether counter-force policy would have a destabilizing effect on the SALT talks, Schlesinger hedged: 'We have no announced counter-force strategy, if by counter-force one infers that one is going to attempt to destroy silos. We have a targeting doctrine that emphasizes selectivity and flexibility' (Schlesinger, 1974c, p. 265). A month later, he made it clear that the USA's targeting policy encompassed 'military' and 'military plus urban/industrial' targets:

> This is not the place to explore the full history and details of that long-standing strategic debate. However, there is one point to note about its results. Although several targeting options, including military and military plus urban/industrial variations, have been a part of U.S. strategic doctrine for quite some time, the concept that has dominated our rhetoric for most of the era since World War II has been massive retaliation against cities, or what is called assured destruction (Schlesinger, 1974d, p. 33).

The motive behind the Administration's rather confused presentation of its policy needs to be disentangled. US targeting policy had always been secret. The declaratory policy of assured destruction had suggested that cities were targeted rather than military installations, and, for various reasons, public officials spoke only about deterrence and city attacks, even though targeting policy encompassed military objectives. When the Nixon Administration came to power there was a genuine concern that massive retaliation was not a credible policy. Limited aggression should, it was felt, be countered in limited ways. A policy of flexible response and controlled escalation seemed to provide an answer to this problem and to hold out the hope that it would be possible 'to end nuclear war quickly if it occurred and to re-establish deterrence as soon as possible' (Lynn Davis, 1975, p. 4). According to Davis, who worked in the Pentagon under Schlesinger, this was the main purpose underlying Schlesinger's flexible response doctrine. To this extent Schlesinger was responding to a concern, expressed in 1973 by Fred Iklé, that 'The jargon of American strategic analysis works like a narcotic. It dulls our sense of moral outrage... It blinds us to the fact that our method for preventing nuclear war rests on a form of warfare universally condemned since the Dark Ages – the mass killing of hostages' (Iklé, 1973, p. 14). Iklé's solution, to make US strategic forces invulnerable and capable of a slow, considered response to aggression, was rejected by others who shared his concern that mutual assured destruction was not a credible basis for policy. For example, Herbert York (1974) advocated 'deep cuts' in force levels as a solution to the problem.

Schlesinger also felt that, now that the Soviet Union had acquired 'the capability in its missile forces to undertake selective attacks against targets other than cities', the USA had to respond: 'if we are to ensure the credibility of our strategic deterrent, to be certain that we have a comparable capability in our strategic systems and in our targeting doctrine, *and to be certain that the USSR has no misunderstanding on this point*' (Schlesinger, 1974d, my italics). Finally, the assumption that deterrence would be improved if the USA had specific plans for a number of contingencies, and if the Soviet Union knew it, was another foundation of Schlesinger's policy (Davis, 1975).

Carter and Reagan: Countervailing Strategy

In common with previous in-coming administrations, the Carter Administration initiated a reappraisal of US strategic policy on 18 February 1977 (Presidential Review Memorandum PRM-10 'Comprehensive Net Assessment and Military Force Posture Review') which was completed towards the end of June and considered by a Cabinet level group chaired by National Security Advisor Zbigniew Brzezinski on 7 July. Reviewing the effects of a nuclear war, the study suggested that the USA would suffer 140 million fatalities against 113 million Russian dead, and that about three-quarters of their respective economies would be destroyed. In such a war, 'neither side could conceivably be described as a winner'. It also concluded that neither side could with advantage to itself launch a limited attack against the other's ICBM force. Finally, it noted that even were the Russians to initiate a first-strike against US strategic forces, the USA would retain clear superiority in bombers and in SLBMs. In the latter field the USA also had a clear superiority in anti-submarine warfare capabilities (Ball, 1983, p. 20).

On the basis of this review President Carter issued Presidential Directive PD-18 'US National Security' on 24 August which, while calling for further studies on ICBM force modernization, maintenance of a strategic reserve force, and nuclear targeting policy, reaffirmed the continued use of National Security Decision Memorandum NSDM-242 and 'Nuclear Weapons Employment Policy' NUWEP-1 in 'the absence of further guidance for structuring the US strategic posture' (House Armed Services Committee, 1977, p. 9), and laid down that the USA should maintain the capability to inflict unacceptable damage on the USSR, even if the USSR had struck first.

The first phase of the review of nuclear targeting policy was completed by December 1978. The second phase remained unfinished when the study group was disbanded in the spring of 1979. The review recommended that command, control, communication and intelligence (C^3I) systems for the control of US nuclear forces should be improved so that they could survive better in a nuclear war, and that more options should be built into the SIOP to give strategic forces 'greater flexibility in targeting than they presently have' (Perry, 1979, pp. 298–9). It also proposed that less emphasis should be placed on the destruction of the Soviet economic and industrial base and greater attention be given to 'improving the effectiveness of our attacks against military targets' (Perry, 1979). As a direct result, work was put in hand on the development of 'a highly complex matrix of targeting "packages" or "building block" options that could be flexibly combined or "tailored" to suit particular situations' (Ball, 1983, p. 21). These options were incorporated into a new 'Nuclear Weapons Employment Policy' document (known as NUWEP-2 or NUWEP-80) issued in October 1980, a revised National Strategic Target List (now listing 40,000 as opposed to 25,000 targets), and a new SIOP which allocated targets to four categories: Soviet strategic nuclear forces; other military forces; Soviet leadership and command and control systems; and industrial and economic targets.

The Nuclear Targeting Policy Review formed the basis for a new Presidential Directive which was drafted in early 1979 but then shelved until it was retrieved just prior to the 1980 Democratic Convention, revised, updated and issued as Presidential Directive PD-59 'Nuclear Weapons Employment Policy' (signed on 25 July 1980). Harold Brown, Carter's Secretary of Defense, was quick to point out (20 August) that PD-59 was 'a refinement, a codification of previous statements of our strategic policy' (Brown, 1980a). Nevertheless it made some changes to US targeting policy. In the area of economic targeting it reflected a shift in emphasis from targeting designed to impede Soviet economic recovery to the targeting of the Soviet economic war-supporting infrastructure. It also reflected the relatively new interest in Soviet political and military leadership targets and military command and control systems (*Boston Globe*, 27 July 1980).

More importantly, PD-59 'emphasised that the pre-planned target packages in the SIOP should be supplemented by the ability to find new targets and destroy them during the course of a nuclear exchange' (Ball, 1983, p. 22). As a result, it emphasized the need to develop new reconnaissance and intelligence systems to provide real-time intelligence capabilities, thus enabling the USA to retarget its forces. Finally, PD-59 required the USA to develop the capability to fight a protracted nuclear war. In one sense there was nothing new about this requirement. NSDM-242 had explicitly directed that the prime subordinate objective of the policy of escalation control was 'maintenance of survivable strategic forces in reserve for protection and coercion during and after major nuclear conflict' (*Washington Post*, 12 October 1980), although the length of time over which these would be needed was not defined; PD-53 (15 November 1979) called for survivable communications in the context of a 'protracted nuclear conflict', while PD-57 'Mobilization Planning' (3 March 1980) emphasized that any war might be quite prolonged (see p. 39).

In the spring of 1981 the Reagan Administration embarked on yet another review of targeting policy. In October it produced National Security Decision Directive NSDD-13 as a successor to PD-59. In July 1982 Secretary of Defense Caspar Weinberger issued a new Nuclear Weapons Employment Policy, known as NUWEP-82. The guidelines in these documents were used to prepare a new SIOP in which increased attention was given to nuclear weapons employment in a situation of prolonged or protracted nuclear conflict.

The current version of the SIOP presents the President with four general attack options. These are major, selected and limited attack options 'designed to permit the selective destruction of fixed enemy military or industrial targets', and regional nuclear options 'intended, for example, to destroy the leading elements of an attacking enemy force' (Ball, 1983, p. 24). The SIOP also gives the President two special attack options: a pre-emptive attack on the Soviet Union; and an all-out retaliation on warning of a nuclear attack (Launch on Warning) or during an actual attack (Launch under Attack). Various sub-options are also identified. 'Reserved' targets including Soviet population centres and national command and control centres are identified. These reserved targets would not be attacked in any of the special options, but could be added to an attack option or attacked at a later stage in the war, should this be felt to be necessary. In addition, the SIOP identifies targets in communist-controlled countries in Eastern Europe and elsewhere.

In his report for Fiscal Year 1980 Carter's Secretary of Defense denied that PD-53, PD-58 and PD-59 were intended to suggest that a nuclear war could be controlled.

> We have no more illusions than our predecessors that a nuclear war could be closely and surgically controlled. There are, of course, great uncertainties about what would happen if nuclear weapons were ever again used... I am not at all persuaded that what started as a demonstration, or even a tightly controlled use of strategic forces for larger purposes, could be kept from escalating to a full-scale thermonuclear exchange (Brown, 1980b, p. 67).

Other members of Carter's Administration also denied that a nuclear war could be won. Walter Slocombe, one of his national security team, was careful to point out that PD-59 'does *not* assume that the United States can win a limited nuclear war, nor does it intend or pretend to enable the US to do so. It *does* seek both to ensure that the United States could prevent the Soviets from being able to win such a war and, most critical, to convince them in advance that they could not win' (cited in Pringle and Arkin, 1983, p. 152). Defence Secretary Harold Brown stated unequivocably: 'Nothing in the policy contemplates that nuclear war can be a deliberate instrument of achieving our national security goals, because it cannot be. But we cannot afford the risk that the Soviet leadership might entertain the illusion that nuclear war could be an option – or its threat a means of coercion – for them' (Brown, 1980a).

Whatever the intention of the Carter Administration, the Reagan Administration went further, specifically proclaiming that the goal of US policy is to prevail in a protracted nuclear war. The classified five-year Defense Guidance Plan issued by Secretary of Defense Caspar Weinberger which was leaked to the *New York Times* and the *Washington Post* (10 November 1982) states that:

> Should deterrence fail and strategic nuclear war with the USSR occur, the United States must prevail and be able to force the Soviet Union to seek earliest termination of hostilities on terms favourable to the United States... US strategic nuclear forces and their command and communications links should be capable of supporting controlled nuclear counter-attacks over a protracted period while maintaining a reserve of nuclear forces sufficient for trans- and post-attack protection and coercion.

During the Administration's first year in office, spokesmen such as Secretary of Defense Caspar Weinberger and Deputy Secretary of Defense Frank Carlucci III called for a US capability to prosecute nuclear war, while others such as Louis O. Guiffrida, head of the Federal Emergency Management Agency and T. K. Jones, Deputy Under Secretary of Defense for Research and Engineering, Strategic and Theatre Nuclear Forces, indicated that a nuclear war could be won. Herbert York, Director of Defence Research and Engineering under President Kennedy, said in an interview: 'What's going on right now is that the crazier analysts have risen to higher positions than is normally the case. They are able to carry their ideas further and higher because the people at the top are simply less well-informed than is normally the case' (Scheer, 1983, p. 13).

Colin Gray is a leading advocate of the nuclear-war-fighting school and a senior arms control adviser to the Reagan Administration. Together with Keith Payne he has argued that 'the United States should plan to defeat the Soviet Union and to do so at a cost that would not prohibit US recovery' (Gray and Payne, 1980, p. 21). They suggested that 'a combination of counter-force offensive targeting, civil defence and ballistic missile and air defence should hold US casualties down to a level compatible with national survival and recovery' (p. 25). Such weapons would not necessarily be used only in retaliation to a Soviet nuclear attack:

> American strategic forces do not exist solely for the purpose of deterring a Soviet nuclear threat or attack against the United States

> itself. Instead, they are intended to support U.S. foreign policy, as reflected, for example, in the commitment to preserve Western Europe against aggression. Such a function requires American strategic forces that would enable a president to *initiate* [my italics] strategic nuclear use for coercive, though politically defensive, purposes (Gray and Payne, 1980, p. 20).

Whereas the concepts of flexible response and escalation control suggest that the Soviet leadership should be spared, at least initially, in order to facilitate intra-war bargaining, Gray and Payne (1980) urged US strategic planners to 'exploit Soviet fears inasfar as is feasible from the Soviet perspective' (p. 15) by having a strategy, based on counter-force targeting, civil defence, air defence and a BMD capability, for destroying the Soviet state while surviving with 'acceptable' casualties of, say, 20 million American dead (pp. 25–6). They argued that 'The most frightening threat to the Soviet Union would be the destruction or serious impairment of its political system.' As a highly centralized society, the Soviet Union would be peculiarly vulnerable to a US attack designed to destroy 'key leadership cadres, their means of communication, and some of the instruments of domestic control'. Under such conditions 'the Soviet Union might cease to function... [and] the USSR might disintegrate into anarchy.' (p. 21). Such a policy 'could have revolutionary consequences for the country' (p. 24).

These ideas have influenced the Reagan Administration. Its acquisition policy is designed to support the drive to achieve a war-fighting and war-winning capability. Richard D. DeLauer, Under Secretary of Defense for Research and Engineering emphasized that the Administration's C^3 programme is the key to Reagan's strategic programme (Scheer, 1983, p. 32). Colin Gray, interviewed by Scheer in May 1982, said: 'The C^3 modernization story doesn't make any sense if you're not thinking along these lines [of protracted nuclear war]... If you only need your forces to go bang on day one, who cares about survivability of the satellites?' (Scheer, 1983, p. 33). Analysing the Administration's acquisition policy, Jeffrey Richelson (1983, p. 141) has pointed out that the capabilities of the weapons and C^3I systems now being acquired by the USA are designed:

1 to destroy time-urgent hardened targets (e.g. MX, Minuteman III with Mark 12A warhead, Pershing II);
2 to deploy more warheads to hit an increasing number of targets and still retain a strategic reserve (e.g. B-1B bomber, air-launched cruise missiles, Trident D-5, etc.);
3 to endure a prolonged nuclear war (e.g. Trident D-5 and various C^3I systems);
4 to provide the flexibility needed to fight a nuclear war (C^3I programmes).

Soviet Nuclear Policy

Surprise Attack

Soviet experience in the Second World War led to the adoption of the following principle as the basis for force planning: 'lots of everything, especially for the army' (Freedman, 1981, p. 59). All types of weapons were thought to have their functions and values. The important thing was not to place undue emphasis on any particular type of weapon or tactic. The American development of the atomic bomb was not discussed in the Soviet military press between 1947 and 1953 (Garthoff, 1958, p. 67) and it was not until after Stalin's death (1953) that the Soviet Union began, in parallel with its acquisition of a stockpile of atomic bombs, to develop a nuclear strategy. Then, with the build-up of its own nuclear forces, it began to reduce the number of men it had under arms and to re-examine its overall military strategy.

One of the first changes was a re-evaluation of the role of surprise attack in war. While this had always been regarded as of tactical importance Stalin, conscious of the effect of the German onslaught on Russia in 1941, had downgraded surprise as a strategic factor lest its consideration led to criticism of his wartime leadership. After his death and the appointment of Marshal Zhukov as Minister of Defence in 1955, the Soviet military began to emphasize the role of pre-emption in modern warfare. In an outspoken article, Marshal of the Tank Forces Rotmistrov wrote: 'It must be plainly said that when atomic and hydrogen weapons are employed, surprise is one of the decisive conditions of the attainment of success not only in battles and operations, but also in war as a whole' (Dinerstein, 1959, p. 186). Rotmistsov suggested that the duty of the Soviet Armed Forces was not only 'to repel the attack successfully but also to deal the enemy counterblows, or even pre-emptive surprise blows, of terrible destructive force' (Rotmistsov, 1955, in Dinerstein, 1959, p. 186).

There is little doubt that by the mid-1950s Soviet leaders were worried by the nuclear superiority of the United States, Eisenhower's massive retaliation policy and the threat of a US nuclear first-strike against the USSR. Defence Minister Nikolai Bulganin went so far as to warn against the danger of a surprise nuclear attack by the USA (*Izvestia*, 22 July 1954), while Khrushchev remarked that Dulles' policy had amounted to 'bare-faced atomic blackmail' (cited in Roberts, 1970, p. 41). There was increasing evidence that Khrushchev accepted that nuclear war would be terrible, but the concept of surprise attack seemed to offer the Soviet Union a way out of the problem by suggesting that it could thwart any US plans for attack by a sudden blow of its own. In 1956 General Krasilnikov argued that 'The successful employment of strategic surprise in the initial phase of the war can lead to disruption of the opponent's existing plans ... An especially intense struggle to pre-empt the opponent will take place in the campaign in the initial phase of the war.' (cited in Freedman, 1981, p. 151). Krasilnikov suggested that the air force might have a crucial role to play. Khrushchev, however, began to speak of bombers as 'obsolete' and to stress the advantages of ICBMs that would be hard to detect on the ground and extremely difficult to counter once launched. In January 1960 he told the Supreme Soviet: 'Military aviation is now being almost entirely replaced by missiles. We have now sharply reduced and probably will further reduce and even halt production of bombers and other obsolete equipment. In the navy, the submarine fleet is assuming greater importance and surface ships can no longer play the role they played in the past' (cited in Warner, 1977, pp. 139–40).

The Rationality of War under Conditions of Assured Destruction

Lenin embraced the Clausewitzian view that 'war is a continuation of policy by other means' and this view was for a long time the Marxist–Leninist concept of war (Garthoff, 1953, pp. 9–19, 51–7).

By the early 1960s, however, Soviet military thinkers accepted that the consequences of a nuclear war would be appalling. G. A. Trofimenko (1976, p. 292) has said that 'the Leninist thesis on the fact that war is a continuation of policy ... by forcible means' is no longer 'in practice a usable instrument of policy when an aggressor in the course of struggle for "victory" can himself be annihilated'.

At the first business meeting of the two SALT negotiating teams at Helsinki (18 November 1969) the Soviet Union, in a prepared statement, said

> Even in the event that one of the sides were the first to be subjected to attack, it would undoubtedly retain the ability to inflict a retaliatory strike of crushing power. Thus, evidently, we all agree that war between our two countries would be disastrous for both sides. And it would be tantamount to suicide for the one who decided to start such a war (cited in Garthoff, 1978, p. 99).

In the publicly available military press mutual deterrence is generally expressed in terms of an assured retaliatory capability which would devastate the aggressor. This deterrent is itself based on a strong military posture which would be used only in the event of war:

> Soviet military power, and the constant enhancement of its capability and readiness, is thus justified primarily for deterrence, as well as to wage a war if one should come despite Soviet efforts to prevent it. This view is consistently held by the Soviet military and political leaders. It is not accurate, as some Western commentators have done, to counterpose Soviet military interest in a 'war-fighting' and hopefully 'war-winning' posture to a deterrent one; the Soviets see the former capability as providing the most credible deterrent, as well as serving as a contingent resort in the event of war (Garthoff, 1978, p. 95).

The Soviet leadership does not appear to believe that it could profitably wage war. As Garthoff points out, professional military writings within the Soviet Union diverge in tone from articles in the *public* military press and clearly stress the mutuality of deterrence. Thus, for example, in the influential (and confidential) military journal *Military Thought*, General Vasendin and Colonel Kuznetzov (1968, p. 42) stated: 'Everyone knows that in contemporary conditions in an armed conflict of adversaries comparatively equal in power (in number and especially in quality of weapons) an immediate retaliatory strike of enormous destructive power is inevitable'. Another article, by then Major General V. I. Zemskov (1969, p. 59), indicated that he believed there was a 'mutual balance' and therefore mutual deterrence. However, he thought that this could be upset by technological advances.

Soviet spokesmen have consistently denied the possibility of initiating a first-strike. Marshal N. I. Krylov (1967, p. 20) noted the existence of 'a system for detecting missile launches' and pointed out that a surprise attack by an aggressor would not succeed because of the capability of the other side to launch a retaliatory attack on warning. At the 1981 UN General Session, Foreign Minister Gromyko submitted a proposal that all countries possessing nuclear weapons should agree not to use them first. He argued that such a statement would lower tensions between the weapons states. The USA and Britain refused. The next year, at the UN Special Session on Disarmament, the Soviet Union declared that it would never use nuclear weapons first. Again, the USA, Britain and France refused to give similar assurances.

At the same time the Soviet leadership has consistently said that any use of nuclear weapons at whatever level will be met by a retaliatory strategic response. Soviet policy is based on the possession of sufficient nuclear forces to survive a US first-strike and retaliate in strength. This does not mean that the Soviet Union will wait for the incoming US, NATO or Chinese missiles to arrive. As soon as there is incontrovertible evidence that an attack has been launched, Soviet strategy calls for a massive nuclear response designed to disrupt the adversary's own attack. At the very least this needs to be interpreted as 'Launch on Warning' (LOW). The semi-official news agency Novosti indicated that Moscow would adopt this policy if NATO deployed Pershing II and cruise missiles in Europe: 'Faced by an infinitesimal warning time, the only possibility remaining is a nuclear retaliatory strike in retribution. There is no other possibility' (*The Guardian*, 31 November 1982). Ball (1981, p. 31) says that 'in the event of a nuclear war, Soviet strategic forces would be used massively rather than sequentially, and against a wide range of nuclear and conventional military targets, command-and-control facilities, centres of political and administrative leadership, economic and industrial facilities, power supplies etc., rather than more selectively.' The notion of 'withheld' targets is foreign to Soviet targeting, so that, for example, US command-and-control facilities would be attacked from the outset (Ball, 1981, p. 32).

There are Soviet theorists who have suggested that a pre-emptive strike might make sense. For example, Colonel-General N. A. Lomov (1973, p. 147) argued that 'One of the decisive conditions for success in an operation is the anticipating of the enemy in making nuclear strikes, particularly against the enemy's nuclear missile weapons.' In practice, Gouré (1974) doubts whether Soviet leaders believe a pre-emptive disarming strike on the West is a practical policy. He believes that while 'Soviet statements appear to envisage that in the event of a threat of war the Soviet Union would attempt to deliver a first counter-force strike there is no indication that the Soviet leaders have a high confidence in being able to do so or that they believe that such a strike would in fact assure the survival of the Soviet Union' (pp. 107–10). What is certain is that Soviet leaders believe 'a defensive strategy based on the concept of absorbing the enemy's first nuclear strike is ... wrong'. Launch on warning is the only way of reconciling this problem.

Limited Nuclear War

Soviet spokesmen reject the idea of limited nuclear war: G. A. Arbatov (1974, pp. 133–4) has argued that

> the idea of introducing rules of the game and of artificial limitations 'by agreement' is based on an illusion and is without foundation. It is hard to imagine that nuclear war, if launched, could be held within the framework of the 'rules' and not grow into general war.

This point was also made by the Soviet Minister of Defence D. F. Ustinov who stated in 1981 that 'None but utterly irresponsible individuals can claim that nuclear war can be fought under some rules established in advance whereby nuclear missiles will go off under a "gentleman's agreement" only at specific targets without hitting the population in the process' (Ustinov, 1981, p. 7).

Western strategists have sought for confirmation that the Soviet Union is seeking to adopt war-fighting strategies similar to those in vogue in the West. However, according to Desmond Ball (1981, p. 33)

> Evidence that Soviet military doctrine now incorporates the possibility of control, selectivity and restraint in a nuclear exchange is actually very fragmentary. It derives principally from some statements that stand aside from the overwhelming thrust of Soviet military literature... Soviet strategic policy and targeting doctrine, together with some quite explicit pronouncements, is to the effect that any nuclear exchange would involve simultaneous and unconstrained attacks on a wide range of targets.

Soviet Targeting Policy

Soviet theory with its emphasis on the totality of nuclear war continues to concentrate on the blitzkrieg, involving the infliction of instantaneous destruction on the enemy. Highest priority is given to the West's nuclear retaliatory force in an effort to limit damage to the Soviet Union. Next come other major military forces, command and control systems, the economic base of society and other targets that support the enemy's capacity to wage war. According to Lomov (1973, pp. 137–8) 'the mass destruction, annihilation or neutralization of these objectives in a short period of time can lead to disorganisation of all vital activities in the enemy nation.'

Soviet Acquisition Policy

Relatively little is known about Soviet doctrine on force size. Soviet military and political leaders ceased to call for strategic superiority as an objective following the 24th Party Congress (April 1971) and instead advocated mutual deterrence, parity and equal security. On the eve of the Carter Administration's accession, Brezhnev disavowed the aim of military superiority:

> Of course, Comrades, we are improving our defences. It cannot be otherwise. We have never neglected and will never neglect the security of our country and the security of our affairs. But the allegations that the Soviet Union is going beyond what is sufficient for defence, that it is striving for superiority in arms, with the aim of delivering a 'first strike' are absurd and utterly unfounded. Our efforts are aimed at preventing both first and second strikes and at preventing nuclear war altogether... The Soviet Union's defence potential must be sufficient to deter anyone from disturbing our peaceful life. Not a course aimed at superiority in arms, but a course aimed at their reduction, at lessening military confrontation – that is

our policy (Radio Moscow, 18 January 1977, *Pravda* and *Izvestia*, 19 January 1977).

G. A. Trofimenko has said that the USSR 'is content with the state of strategic parity and does not strive for superiority' (Trofimenko, 1981, p. 39). Although US analysts tend to dismiss such statements, Raymond Garthoff concluded that the record 'indicates that the Soviet political and military leadership accepts a strategic nuclear balance between the Soviet Union and the United States as a fact, and as the probable and desirable prospect for the foreseeable future' (Garthoff, 1978, p. 120). The importance of maintaining this balance was stressed by Brezhnev on 2 November 1977 on Radio Moscow, and repeated in *Pravda* and *Izvestia* (3 November 1977):

> The Soviet Union is effectively looking after its own defences, but it does not and will not seek military superiority over the other side. We do not want to upset the approximate balance of military strength existing at present... between the USSR and the United States. But in return we insist that no one else should seek to upset it in his favour.'

Soviet leaders have expressed concern at the US build-up of strategic arms and capabilities. Brezhnev (1970, p. 541) said: 'We shall respond to any and all attempts from any quarter to obtain military superiority over the USSR with a suitable increase in military strength to guarantee our defence. We cannot do otherwise'. A year earlier General Zemskov (1969) had argued that any 'disruption' of the balance of power in which the West achieved an advantage would 'increase greatly the danger of a nuclear war' (p. 59). The two factors which Zemskov identified as potentially destabilizing were the development of an ABM capability or a 'further sharp increase of nuclear [strike] potential' (p. 59). More recently, Colonel Semeyko (1979) has stated that 'The correlation of strategic forces is very sensitive to disruption. There could be an equal number of means of delivering strategic nuclear weapons, but if one side insured superiority for itself in all or nearly all of the most important characteristics of those weapons, it would possess superiority'. This destabilization is now occurring.

Current American procurement and deployment programmes, including MX, Trident, Pershing II, Cruise and B-1 bomber, have been condemned by Soviet commentators as 'clearly revealing Washington's course directed at overthrowing the existing approximate balance of forces between the USSR and the United States and at achieving American military superiority' (*Pravda*, Editorial, 29 January 1980). Prime Minister Alexsei Kosygin, speaking on Radio Moscow on 21 February 1980, stated that 'no one must be left in any doubt that the Soviet Union will not allow any disruption of the balance of forces which has come about in the world to the detriment of its security.' Given this perception, it is not surprising that the Soviet Union has announced its intention to increase its own strategic capabilities.

Conclusions

Emphasis on the doctrinal shifts in US strategic nuclear policy can obscure the underlying consistency in US employment policy as laid down in nuclear war plans. The USA has consistently planned to use nuclear weapons against military targets as well as against counter-value targets. In assessing US strategic doctrine three factors are particularly important:

1 *First-use*. The USA has always been prepared to use nuclear weapons first. *First-use* should not be confused with *first-strike*. However, whereas the USA believes that first-use will not necessarily escalate into all-out nuclear war, the Soviet Union draws no such distinction.

2 *Mutual Assured Destruction*. Clearly if you believe that you will be destroyed in a nuclear war, whoever starts it, there is an incentive not to start one. This idea lay at the heart of deterrence theory. However, deterrence is not a stable condition and it could be upset by technological advances in offensive and defensive weapons design that encourage one side to believe that it could fight and win a nuclear war. This is now happening. It is in this respect that the 'nuclear winter' findings discussed in chapter 6 (pp. 148–50) are so important. They impose a supra-human 'sword of Damocles' over the whole question of 'getting away' with initiating a nuclear war.

3 *Nuclear war-fighting*. The USA could clearly have waged a nuclear war against the Soviet Union in the late 1940s and early 1950s and emerged relatively unscathed. By the late 1950s this

was no longer a credible option. For a while thought was given to the option of limited nuclear war and indeed the USA retains this within its employment policy. However, it is an option that the Soviet Union has consistently rejected out of hand. The USA is now trying to re-establish nuclear war-fighting and war-winning as a credible option through the acquisition of counter-force and BMD weapons. Whether or not the USA would use its power to wage a nuclear war is an open question. It clearly feels that having the capability is politically important – a point taken up in chapter 5 (pp. 116–18).

It is much less easy to analyse Soviet declaratory and employment doctrine. Soviet nuclear doctrine is generally not as openly expressed as is that of the USA and Soviet thinking has to be deduced from very general statements by political and other leaders and from military writings and force dispositions. Robert Legvold (1979, p. 9) went so far as to suggest that the essential difference between the US and Soviet positions is that, while the former tries to have a strategic doctrine, the latter does not: 'Soviet observers regard [strategic nuclear doctrines] as ill-considered attempts to rationalise the use of nuclear weapons'. The USSR, Legvold believes, concentrates on operational concepts of war.

Western analysis of Soviet military writings tends to move between the two poles of those who believe that the Soviets adhere to Marxist–Leninist doctrine, believe they can fight and win a nuclear war, and are acquiring weapons to do this; and those who believe that Soviet military writings sustain ideological positions and military morale rather than reflect Soviet leaders' true opinions. The latter point out that it is the duty of military leaders to plan to fight and win a war, should one occur, but that there is no evidence that Soviet political leaders are planning nuclear war. As Brezhnev put it, 'anyone who starts a nuclear war in the hope of winning it has thereby decided to commit suicide' (*Pravda*, 1 October 1981).

From a European point of view, it is significant that there are two US scenarios, one for strategic nuclear policy, the other for theatre nuclear policy. At least from the early 1950s the USA has believed that nuclear weapons might be used in Europe (and in other theatres) in a controlled and limited way, which would not necessarily result in a strategic nuclear exchange between the superpowers. There are broadly speaking two schools of thought on this: those who believe that a limited nuclear war will escalate to all-out nuclear war very quickly; and those who believe that it might be contained. President Reagan, for example, was reported in the *International Herald Tribune* (21 October 1981) as saying: 'I could see where you could have an exchange of tactical weapons against troops in the field without it bringing either one of the major powers to pushing the button.' Such a policy may neatly divide US strategic and theatre nuclear policy and be thought to protect the USA from the likely effects of nuclear war by providing it with the chance to win a theatre war. In Europe such a war could not be won. Everyone would lose.

Soviet leaders reject out of hand any conception that a nuclear war could be limited. Speaking in East Germany on 6 April 1983, the Soviet Defence Minister Marshal D. F. Ustinov stated flatly (1983): 'If they think in Washington that we will retaliate [to US use of European Forward Based Systems] only against targets in Western Europe, they badly delude themselves. Retaliation against the United States will be ineluctable'.

References and Further Reading

Arbatov, G. A. (1974) in *Survival*, 16(3), cited in Openshaw, S., Steadman, P. and Greene, O. (1983) *Doomsday. Britain after Nuclear Attack*, Oxford: Basil Blackwell.

Ball, D. (1981) *Can Nuclear War be Controlled?* London: International Institute for Strategic Studies.

Ball, D. (1983) *Targeting for Strategic Deterrence*, London: International Institute for Strategic Studies.

Bell, C. (1957) 'Atoms and Strategy', in F.C . Fesham (ed.) (1957) *Survey of International Affairs 1954*, London: Oxford University Press.

Brennan, D. (1961) *Arms Control, Disarmament and National Security*, New York: George Braziller.

Brezhnev, L. I. (1970) *Leninskim Kursom*, Moscow: Politizdat, vol. 3.

Brodie, B. (1959) *Strategy in the Missile Age*, Princeton: Princeton University Press.

Brodie, B. (1973) *War and Politics*, New York: Macmillan.

Brown, H. (1980a) 'The objective of US strategic forces'. Address given to the Naval War College, 22 August 1980, Washington DC: US International Communication Agency.

Brown, H. (1980b) *Annual Department of Defense Report, Fiscal Year 1980*, Washington DC: Government Printing Office.

Davis, L. E. (1975) *Limited Nuclear Options: Deterrence and the new American Doctrine*, Adelphi Papers No. 121, London: International Institute of Strategic Studies.

Dinerstein, H. S. (1959) *War and the Soviet Union*, New York: Praeger.

Dulles, John Foster (1954a) 'The Evolution of Foreign Policy', *US Department of State Bulletin*, 30, 107–10.

Dulles, John Foster (1954b) 'Policy for Security and Peace', *Foreign Affairs*, 32, 353–64.

Freedman, L. (1981) *The Evolution of Nuclear Strategy*, London: Macmillan.

Garthoff, R. L. (1953) *Soviet Military Doctrine*, Glencoe, Illinois: Free Press.

Garthoff, R. (1958) *Soviet Strategy in the Nuclear Age*, New York: Praeger.

Garthoff, R. (1978) 'Mutual deterrence, parity and strategic arms limitation in Soviet policy', in D. Leebaert (ed.) (1981) *Soviet Military Thought*, London: George Allen & Unwin Ltd.

Gouré, L. et al. (1974) *The Role of Nuclear Forces in Current Soviet Strategy*, Coral Gables, Florida: Center for Advanced International Studies.

Gray, C. S. and Payne, K. (1980) 'Victory is Possible', *Foreign Policy*, 39, 14–27.

House Armed Services Committee (1977) *Hearings on Military Posture and H. R.* 1872, Washington DC: US Government Printing Office, Book 1 of Part 3.

Iklé, F. C. (1973) 'Can nuclear deterrence last out the century', *Foreign Affairs*, 51, 2; also in R. Pranger and R. Labrie (eds.) (1977) *Nuclear Strategy and National Security: Points of View*, Washington, DC: American Enterprise Institute.

Kaufmann, W. (1956) *Military Policy and National Security*, Princeton: Princeton University Press.

Kaufmann, W. (1964) *The McNamara Strategy*, New York: Harper and Row.

Kennedy, J. F. K. (1962) cited in *Newsweek*, 9 April.

King, J. (1957) 'Nuclear plenty and limited war', *Foreign Affairs*, 35, 238–56.

Kissinger, H. A. (1955) 'Military Policy; Defense of the "Grey Areas" ', *Foreign Affairs*, 33, 416–28.

Kissinger, H. A. (1960) *The Necessity for Choice*, London: Chatto and Windus.

Kistiakowsky, G. B. (1976) *A Scientist at the White House*, Cambridge: Harvard University Press.

Krylov, Marshal N. I.(1967) 'The nuclear missile shield of the Soviet State', *Voyennaya mysl'*, (*Military Thought*), November.

Legvold, R. (1979) 'Strategic Doctrine and SALT: Soviet and American Views', *Survival*, 21 (1), 8–13.

Lomov, Colonel-General N.A. (1973) *Scientific-technical progress and the Revolution in Military Affairs*, Moscow. Translated by the USAF and published in *Soviet Military Thought*, Washington DC: US Government Printing Office, Vol. 3.

McNamara, R. S. (1963a) *Statement on the Fiscal Years 1964–1968 Defense Program and 1964 Defense Budget* (27 January), Washington, DC: US Government Printing Office.

McNamara, R. S. (1963b) *Department of Defense News Release* 1486–63, 18 November.

McNamara, R. S. (1965)) *Statement on the Fiscal Years 1966–1970 Defense Program and 1966 Defense Budget* (18 February), Washington, DC: US Government Printing Office.

McNamara, R. S. (1967) *Statement on the Fiscal Years 1968–1972 Defense Program and Defense Budget for Fiscal Year 1968* (23 January), Washington DC: US Government Printing Office.

McNamara, R. S. (1968) *Statement on the Fiscal Years 1969–1973 Defense Program and Defense Budget for Fiscal Year 1969* (22 January), Washington DC: US Government Printing Office.

McNamara, R. S. (1983) 'The military role of nuclear weapons: perceptions and misperceptions', *Foreign Affairs*, 62 (1), 59–80.

National Security Council:

NSC-30 (1948) 'United States Policy on Atomic Warfare', US Department of State, Foreign Relations of the United States, 1948, Vol. 1: *The United Nations*, Washington, DC: US Government Printing Office, 1976, 624–8.

NSC-68 (1950) 'United States Objectives and Programs for National Security' (14 April), US Department of State, Foreign Relations of the United States, 1950, Vol. 1: *National Security Affairs: Foreign Economic Policy*, Washington, DC: US Government Printing Office.

NSC-162/2 (1953) Review of Basic National Security Policy (30 October), NSC Papers, Modern Military Branch, National Archives, Washington DC.

NSC-5540/1 (1954) 28 December, approved as Basic National Security Policy in NSC-5501 (6 January 1955), paragraph 35.

NSC-5602/1 (1956) 'Basic National Security Policy' (15 March), Basic National Security Policy Folder, NSC Series, Policy Papers Series, Box 17, WHO-SANSA, DDEL, 1–11.

National Security Decision Directive:

NSDD-13 (1981) *Nuclear Weapons Employment Policy*, classified.

National Security Decision Memorandum:
NSDM-242 (1974) *Policy Guidance for the Employment of Nuclear Weapons*, classified.
NATO (1968) *Overall Strategic Concept for the Defence of the NATO Area*, Military Committee MC 14/3, 16 January 1968.
Nixon, R. (1970) *A Report to the Congress: U.S. Foreign Policy for the 1970s. A New Strategy for Peace* (18 February), Washington DC: US Government Printing Office.
Nixon, R. (1971) *A Report to the Congress: U.S. Foreign Policy for the 1970s. Building for Peace* (25 February), Washington DC: US Government Printing Office.
Nuclear Weapons Employment Policy, classified documents:
NUWEP-1 (1974).
NUWEP-2 (1980), also known as NUWEP-80.
NUWEP-82 (1982).
Perry, W. J. (1979) *Testimony to the Senate Armed Services Committee, Department of Defense Authorization for Appropriations for Fiscal Year 1980*, Washington DC: US Government Printing Office, part 1.
Presidential Directives, classified documents:
PD-18 (1977) *US National Security*.
PD-53 (1979) *National Security Telecommunications Policy*.
PD-57 (1980) *Mobilization Planning*.
PD-58 (1950) *Continuity of Government*.
PD-59 (1980) *Nuclear Weapons Employment Policy*.
Presidential Review Memorandum:
PRM-10 (1977) *Comprehensive Net Assessment and Military Force Posture Review*.
Pringle, P. and Arkin, W. (1983) *SIOP. Nuclear War from the Inside*, London: Sphere Books.
Richelson, J. (1983) 'PD-59, NSDD-13 and Reagan's Strategic Modernisation Program', *Journal of Strategic Studies*, 6 (2), 125–46.
Roberts, C. (1970) *The Nuclear Years: the Arms Race and Arms Control 1945–70*, New York: Praeger.
Rosenberg, D. A. (1983) 'The Origins of Overkill. Nuclear Weapons and American Strategy, 1945–1960', *International Security*, 7 (4), 3–71.
Rowen, H. S. (1975) 'Formulating Strategic Doctrine', in Commission on the Organization of the Government for the Conduct of Foreign Policy (1975), Washington, DC: US Government Printing Office, vol. 4, appendix K, *Adequacy of Current Organization: Defense and Arms Control*.
Scheer, R. (1983) *With Enough Shovels. Reagan, Bush and Nuclear War*, London: Secker and Warburg.
Schlesinger, J. R. (1974a) *Address to the Overseas Writers Association*,

Washington DC, 10 January. Excerpts published in 'Flexible Strategic Options and Deterrence', *Survival*, 16 (2), 86–90.
Schlesinger, J. R. (1974b) Press Conference, 23 January.
Schlesinger, J. R. (1974c) *Fiscal Year Authorizations for Military Procurement, Research and Development, and Active Duty, Selected Reserve and Civilian Personnel Strengths*, Hearings before the Senate Armed Services Committee, 5 February, part 1.
Schlesinger, J. R. (1974d) *Annual Defense Department Report, Fiscal Year 1975*. Secretary of Defense (4 March), Washington DC: US Government Printing Office.
Semeyko, Colonel L. (1970) 'Imperialism's Strategic Concepts: the Course Toward Military Superiority', *Krasnaya zvezda*, 24 March 1979.
Sloss, L. (1981) 'The Evolution of the Countervailing Strategy', mimeo, May 1981, cited in D. Ball (1983) *Targeting for Strategic Defence*, London: International Institute for Strategic Studies.
Trofimenko, G. A. (1976) *SSLA: Politika, voina, ideologiya, (The USA: Politics, War and Ideology)*, Moscow: Mysl'.
Trofimenko, G.A. (1981) 'Counterforce: illusion of a panacea', *International Security*, 5 (2), 142–58.
Ustinov, D. F. (1981) *Against the Arms Race and the Threat of War*, Moscow: Novosti Press.
Ustinov, D. F. (1983) cited in *The Guardian*, 7 April.
Vasendin, Major General N. and Kuznetzov, Colonel N. (1968) 'Contemporary War and Surprise', *Voyennaya mysl'*, (*Military Thought*), June.
Warner, E. L. (1977) *The Military in Contemporary Soviet Politics*, New York: Praeger.
York, H. (1974) 'Reducing the Overkill', *Survival*, 16, 2.
Zemskov, Major General V. I. (1969) 'Wars of the Contemporary Era', *Voyennaya mysl'*, (*Military Thought*), May.

4
The British Independent Nuclear Deterrent

Britain was the first country to try to build an atomic bomb. In view of the war development work was transferred to the USA on the grounds that the project would be safer from attack if it were located in the USA and also because the Americans could put in far more resources than the British (see chapter 1). Early in 1945, however, a scientific committee chaired by Sir Henry Tizard concluded that the only defence against an enemy armed with atomic bombs would be to have one's own and to threaten to use them in retaliation. By October the Chiefs of Staff had come to the same conclusion and in January 1946 they indicated that they were 'convinced we should aim to have as soon as possible a stock [of atomic bombs] in the order of hundreds rather than scores' (cited in Gowing, 1974, p. 169). It was a view with which Prime Minister Clement Attlee agreed.

By the summer of 1946 the decision to make atomic weapons had been implicitly taken when it was decided to build an atomic pile capable of producing sufficient plutonium for about 15 bombs a year. By July that year the Chiefs of Staff were arguing for the development of a high-performance long-range bomber to deliver the bombs. Formally, however, the decision to build a bomb was not taken until early January 1947 when an *ad hoc* committee of the Prime Minister and six of his colleagues decided, in great secrecy, to go ahead. Of this decision Margaret Gowing commented:

> The British decision to make an atomic bomb had "emerged" from a body of general assumptions. It had not been a response to an immediate military threat but rather something fundamentalist and

> almost instinctive – a feeling that Britain must possess so climactic a weapon in order to deter an atomically armed enemy, a feeling that Britain as a great power must acquire all major new weapons, a feeling that atomic weapons were a manifestation of the scientific and technological superiority on which Britain's strength, so deficient if measured in sheer numbers of men, must depend (Gowing, 1974, p. 184).

It was also, as Gowing notes, 'a symbol of independence', a belief that reflected Ernest Bevin's view, expressed to the *ad hoc* committee, that 'We could not afford to acquiesce in an American monopoly of this new development' (cited in Gowing, 1974, p. 183). Public confirmation of the decision to build a British bomb was given in Parliament on 12 May 1948 in response to a parliamentary question.

While Britain's commitment to nuclear weapons appeared to confirm the country's continuing claim to status as a great power, as early as 1949 it was clear the drive to develop an atomic bomb was diverting research and development funds from other projects, including aircraft capable of delivering the bombs to their target. Faced with conflicting demands and a limited budget, the Chiefs of Staff decided in the summer of 1949 that 'for political and strategic reasons' it was essential to press on with the production of atomic energy and atomic weapons in Britain. Nevertheless, research and development of the atomic weapon and the means of delivering it, while of the highest priority, should be considered alongside other vital defence research and development projects (Gowing, 1974, p. 227).

The Soviet explosion of an atomic bomb led to a further review of Britain's atomic weapons programme. Although Sir Henry Tizard challenged the view that Britain should develop an independent strategic atomic air offensive capability, on the grounds that 'it was foolish for Britain to do things that America could do so much better' (Gowing, 1974, p. 230), the prevailing view was that Britain, as leader of Western Europe and the Commonwealth, should develop a deterrent. Reflecting these views, the Chiefs of Staff recommended in April 1950 that Britain should continue to develop nuclear weapons and build up a small stockpile of its own, a view with which the Prime Minister and the Foreign Secretary concurred. In December 1950, the first production order was placed for 25 Valiant bombers to carry Britain's atomic bomb.

The Conservative Government under Winston Churchill, like its Labour predecessor, favoured the development of nuclear weapons. In a world dominated by the rhetoric of the first Cold War, the Government was concerned at the soaring costs of conventional defence and particularly by the implications of proposals to strengthen NATO conventional forces (see p. 47). Believing that no country would dare attack a nuclear power, Churchill urged the build-up of a nuclear deterrent. In 1952 firm orders were placed for the V-bomber fleet at the expense of a cut in the rate of production of Canberras, an aircraft more suited to the NATO role. The first production aircraft was delivered to the Royal Air Force (RAF) in November 1953, and the first aircraft entered service in January 1955, becoming operational towards the end of 1956.

During 1952 work also began on the British hydrogen bomb project although it was only made public in the 1955 Defence White Paper (Cmnd. 9391). This White Paper also promised the development of a new nuclear weapon delivery system – Blue Streak and enunciated the new policy of massive retaliation. Meanwhile, conventional forces were to be cut, a move that had been anticipated in the 1954 Defence White Paper (Cmnd. 9075, p. 5) which had noted that 'still greater emphasis' was to be placed on the strategic role of the RAF, while the size of the army was to be reduced gradually. The cuts in conventional forces continued under the leadership of Sir Anthony Eden who initiated a thorough-going defence review in the summer of 1956. By mid-July Eden was proposing to President Eisenhower the retention of a 'shield' of conventional forces in Europe but arguing that this 'is no longer our principal military protection' (Eden, 1960, p. 371).

At this stage the main rationale for an independent British nuclear force rested on three principles. Firstly, unless Britain had its own nuclear forces, it could not be sure 'that in an emergency the resources of the other powers would be planned exactly as we would wish, or that the targets which would threaten us most would be given what we consider the necessary priority or the deserved priority, in the last few hours' (Churchill, 1955). In other words, an independent nuclear deterrent gave Britain independent targeting options. Secondly, Britain (and Europe) could not be defended except through deterrence. As a direct result Britain 'must, in our allocation of resources, assign even higher priority to the primary deterrent' (Cmnd. 9391, 1955). Thirdly, an independent nuclear force was necessary for political influence and to assert independence *vis-à-vis* the USA. For this reason, the Minister of Defence, Harold Macmillan, rejected the notion of total dependence on the USA on the grounds that 'this is a very dangerous doctrine... Politically it surrenders our power to influence American policy and then, strategically and tactically, it equally deprives us of any influence over the selection of targets and the use of our vital striking forces. The one, therefore, weakens our prestige and our influence in the world, and the other might imperil our safety' (Macmillan, 1955).

When Macmillan became Prime Minister in 1957 his Minister of Defence, Duncan Sandys, prepared a Defence White Paper that aimed to reduce military manpower drastically from 690,000 to 375,000 and to place more emphasis on deterrent forces while leaving the range of defence obligations unchanged. The role of the V-bomber force, which had become operational at the end of 1956, was confirmed and the decision that it would be 'supplemented by ballistic rockets' (Blue Streak) was announced (Cmnd. 124, p. 9). RAF fighter command aircraft were to be substantially cut and replaced by ground-to-air missiles. National conscription was to be phased out. RAF transport command was to be strengthened to enable Britain to respond to crises overseas. A reduction in the British Army of the Rhine was also announced (and this was followed by a further reduction in January 1958). Defence policy was to be based on 'massive retaliation' and 'defence with fewer men'. The policy was condemned by Labour's Richard Crossman as 'a reckless gamble – a gamble on the nuclear deterrent and a gamble on regular recruitment' (cited in Groom, 1974, p. 213). Crossman argued that the deterrent should be left to the Americans: 'By first adding a few bombs marked "B for Britain" to the bombs marked "A for America" we do not alter in any way the balance of power.' Britain, he said, should relinquish its independent force otherwise 'How can we possibly prevent the Germans, the French and every other nation in the alliance saying "What the British demand for themselves we demand for ourselves"? The right to distrust the Americans cannot remain a British monopoly' (Crossman, 1958).

During 1957 Macmillan worked hard to repair Anglo-American relations which had suffered as a result of the 1956 Suez crisis. The development of a British deterrent, in part as an insurance against renewed American isolationism, persuaded the Americans that Britain was a fully fledged nuclear power, and Macmillan's most notable success was in restoring Anglo-American cooperation in the field of strategic weapons. During February 1958 Macmillan agreed to the stationing of 60 American Thor missiles in Britain under a two-key system of control (see p. 14). A subsequent agreement in November 1960 approved the establishment of a US Polaris Submarine base at Holy Loch. The same period saw increased coordination on an informal basis between Bomber Command and the US Strategic Air Command, and the integration of their nuclear striking forces through joint targeting procedures. This collaboration was reflected in the Anglo-American agreement of 1959 which allowed Britain to buy from the USA 'component parts of nuclear weapons and weapons systems and to make possible the exchange of British plutonium for American enriched uranium'. Although there was a two-way flow of information, there was no doubt that Britain was becoming increasingly dependent on the USA in some areas of defence. This was illustrated by the problems encountered in planning the replacement of the V-bomber force.

The Defence White Paper of 1958 (Cmnd. 363) had confirmed Britain's dependence on nuclear deterrence and stated that while the Western nations would never start a war against the USSR 'it must be well understood that, if Russia were to launch a major attack on them, even with conventional forces only, they would have to hit back with strategic nuclear weapons.' The effectiveness of the deterrent clearly depended on the viability of its delivery system. In 1958 it could still be claimed that the British deterrent was fully independent, but this situation was about to change. In the mid-1950s the British Government had begun to develop the liquid-fuelled Blue Streak rocket to succeed the V-bomber force. Soviet advances, however, made it obsolete before it could be deployed since, with only 4 minutes warning of a Soviet first-strike, Blue Streak could only be used as a first-strike weapon. In April 1960 the Conservative Government acknowledged these weaknesses and scrapped the military version of the missile.

It was becoming clear that development of an independent British delivery system was very expensive yet the Government was unwilling to admit that its defence policy needed to be rethought yet again, particularly since there was opposition to it from the Labour Party and the Campaign for Nuclear Disarmament. It therefore decided to purchase a delivery system from the USA. The choice initially fell on Skybolt, an American built air-launched missile which was to have a British warhead and be fitted to the V-bombers, thus extending their useful life. The Skybolt agreement included a clause binding the British to joint consultation with the USA before the missile was used.

Already embarrassed by its cancellation of Blue Streak, the Government bought Skybolt only to find, at the end of 1962, that the US Government was going to scrap the project in favour of the submarine-launched Polaris. At a hastily arranged meeting with President Kennedy, Prime Minister Macmillan accepted the American offer of Polaris (less warheads) as a replacement for use in British submarines. The Nassau agreement pledged Britain to assign the submarines 'as part of a NATO force and targeted in accordance with NATO plans' while retaining a residual right, which the British argued maintained their independence, to use the missiles in cases 'where Her Majesty's Government may decide that supreme national interests are at stake'. The two countries also agreed on the importance of increasing the effectiveness of their conventional forces (Cmnd. 1915, 1962). This reflected the 'graduated and flexible response' policy which had been developed by the American Secretary of Defense, Robert McNamara, to deal with the new strategic situation in which the two superpowers had the capability for mutual assured destruction (see pp. 53–5).

In many ways the Nassau Agreement settled British defence policy for the next 15 years against a background in which the importance of the Anglo-American 'special' relationship declined. Initially the 1964–70 Labour Government, while recognizing that Britain's defence commitments were 'seriously overstretched' (1965 Defence White Paper, Cmnd. 2592) maintained a commitment to a world role. The extent of over-commitment was demonstrated by the Government's inability to deal militarily with the Rhodesian Unilateral Declaration of Independence in November 1965. Then the economic crisis of 1966–8 forced a fundamental review of British defence thinking which was to be implemented in the following years and which led, under the Conservative Government of Edward Heath (1970–4) to the development of a Euro-centric defence policy. Indeed, since the Second World War, the major shift in British defence policy has been from pursuing a global role to a positive commitment to NATO within Europe.

So far as the deterrent was concerned Labour's policy was essentially pragmatic. The Government argued that the idea of an 'independent' British deterrent which Macmillan had written into the Nassau Agreement was a pretence. The V-bomber force was assigned to NATO as a contribution to its strategic nuclear deterrent. The Polaris force, when acquired, would be technologically dependent on the USA. Nevertheless, the Defence Secretary, Denis Healey, believed that it provided a 'second centre of decision for the first use of nuclear weapons'. As such it would allow Britain to initiate the use of nuclear weapons even if the USA held back, and was an insurance against the failure of the American nuclear guarantee. In line with these views the Labour Government of 1964–70 (and subsequent ones) have emphasized the contribution which Britain's deterrent makes to the alliance as a whole. This view was expressed by Prime Minister Harold Wilson in the debate on the 1966 Defence Estimates (Cmnd. 2901), when he remarked: 'We intend to collectivise our nuclear position whether in the Atlantic, Pacific or anywhere else... it would be part of a collective security agreement with no pretence at independent national status' (Wilson, 1966).

The Conservative Government of 1970 maintained the view that NATO and European security demanded both nuclear and conventional forces. The British deterrent was now firmly based on Polaris although the V-bomber force remained in existence and Britain also had tactical nuclear forces deployed in Germany, though subject to NATO guidelines on their use. Some thought was given to the modernization of the Polaris force. The possibility of converting it to take the American Poseidon submarine launched ballistic missile (SLBM) was considered. The Government went as far as asking the Nixon Administration if Poseidon was available. Kissinger, however, believing that this would jeopardize the US-Soviet arms limitation talks, advised Nixon to refuse and the British were in effect told 'Don't ask' (*The Sunday Times*, 7

April 1985). As a direct result Harold Wilson, who while in office in 1964–70 had initiated work on a British Polaris-improvement programme, decided upon his return to office in 1974 to go ahead with the proposed Chevaline warhead system. Chevaline, which was approved in November 1974, was intended to enable the warheads to penetrate the Soviet anti-ballistic missile defences around Moscow by providing a MRV capability. It is not known how many warheads are delivered by each missile. It is generally assumed that there are three, but *The Sunday Times* Insight Team believes that it is only two (*The Sunday Times*, 7 April 1985).

By the late 1970s the question of a replacement for the Polaris boats was being raised. The real problem was the age of the four submarines. When they were constructed it was known they would have an active life of 20 years although subsequent statements mentioned the figure of 25 years. They were brought into service between 1967 and 1969, so their active life should come to an end around 1990. The submarines' operating lives can be extended by reducing the depth at which they patrol, but although this reduces wear on the hulls, there remains the problem of the build-up of background radiation from their nuclear reactors. Increasing age also increases maintenance and refit costs. Since four submarines is the minimum number needed to ensure that at least one is on patrol at all times, the deterrent force will lose much of its credibility as soon as the first is withdrawn.

Publicly the Labour Government gave the impression that it was not considering the issue. Defence Secretary Fred Mulley explained that 'the existing Polaris fleet will be effective for many years and that being the case, there is no need to take a decision on whether any other arrangements would have to be made' (Mulley, 1978). Not until May 1979 did a Parliamentary Committee begin to investigate the problem of replacing Polaris. Actually, a small *ad hoc* Cabinet group (Prime Minister James Callaghan, Chancellor Denis Healey, Foreign Secretary David Owen and Defence Secretary Fred Mulley) had already met during 1977 and set in train a review of possible replacements for Polaris. By the end of 1978, although no firm conclusions had been reached, it was generally agreed that the optimum time for a decision would be towards the end of 1980, and that a replacement should probably be a submarine. Subsequently, Callaghan, who was pro-Trident, discussed the question with President Carter when they met in

Guadeloupe in December 1978: 'Carter promptly indicated that Britain could have Trident if it wanted it. He suggested that Callaghan put the request in writing' (*The Sunday Times*, 7 April 1985). No letter was, however, sent.

The May 1979 General Election brought a Conservative Government to power. Committed to the nuclear deterrent, the Cabinet established a subcommittee consisting of Prime Minister Margaret Thatcher, Home Secretary William Whitelaw, Chancellor Geoffrey Howe, Defence Secretary Francis Pym and Foreign Secretary Lord Carrington, to consider the replacement for Polaris. Benefitting from the work done under Labour, opinion confirmed the preference for a submarine-based system. During the latter half of 1979 there were signs that the British would once again turn to the USA for a replacement.

A communiqué issued after Mrs Thatcher's December 1979 meeting with President Carter noted agreements 'on the importance of maintaining a credible British strategic deterrent and US/UK strategic co-operation', and that talks should continue on 'the most appropriate means of achieving these objectives for the future' (Joint Statement, 1979). In January 1980 Parliament debated the nuclear force. Criticism from the Labour opposition concentrated on the cost and budgetary impact of a new force, and not the advisability or otherwise of having it. The broad consensus between the two major parties on nuclear policy was confirmed, and in July 1980 the Government announced that it would be procuring Trident, although there was some doubt whether Britain would deploy the Trident Mark I (C-4) which entered US service in July 1980, or Trident Mark II (D-5) which will enter service some time in the latter half of the 1980s. The choice was effectively made for the Government by the Americans, who, having agreed to provide Britain with the existing Trident C-4, then made it clear that the C-4 was going to be phased out and replaced by the more sophisticated and expensive Trident D-5 version (*The Sunday Times*, 7 April 1985). Early in 1982 the Government announced that it was opting for the more advanced Trident D-5 missile which would require even bigger submarines. The D-5 has the capability to carry up to 17 warheads, although the USA is likely to deploy it with 7 or 8. Implementation of the D-5 programme would provide Britain with perhaps 512 but possibly more independently targetable warheads, against its present 64. No direct conditions were attached to the purchase although one American party to the deal said that 'the attitude of the administration, and even more that of Congress, is conditioned by the way the UK acts in the wider defence interests of the alliance and the US' (*The Sunday Times*, 21 February 1982).

The decision to acquire Trident II, defended by the Government on the grounds that it preserved 'commonality' with the Americans, has been opposed by the Labour, Liberal and Social Democratic parties while the high cost of the programme (now over £10 billion) could yet lead to its cancellation. If such action were taken there are a number of options available which could be taken to maintain a British nuclear deterrent. These include the deployment of American sea- or air-launched cruise missiles, and the collaborative development of a new intermediate-range missile based on Pershing.

The Government (Cmnd. 9430, 1985) accepts that 'Trident will undoubtedly cost a lot of money' but insists that it 'will consume only a small part of the real increase in the defence budget since 1978–79'. It argues that, 'while the capital cost of a strategic nuclear deterrent force is considerable, running costs are comparatively small: Polaris, for instance, accounts for less than 2% of the defence budget' (p. 7). Thus 'impressions that we could sustain much larger conventional forces without a Polaris replacement than with it are, therefore, well wide of the mark' (p. 8). Trident is certainly affordable within the context of the defence budget as a whole but, with half the equipment budget going on maintenance and spares, it will absorb more than 20 per cent of the cash available for new equipment in its peak years. With the phasing of expenditure on Trident firmly fixed, it is other major equipment purchases that will suffer.

Critics argue that Trident, with its increased accuracy, is a 'first-strike' weapon. The Government denies this:

> A first strike would involve a surprise attack intended to destroy the other side's retaliatory capability before it could be launched. Such a concept plays no part in the thinking of NATO, the United States, or the United Kingdom: our policy is solely to deter an attack. … We neither have nor seek to have a first-strike capability; and we could not achieve it even if we wished (Cmnd. 9430, 1985, p. 7).

Yet, as chapter 3 showed (pp. 61–4), American strategic theory is now greatly influenced by plans to wage nuclear war including 'decapitating' attacks aimed at paralyzing and disrupting the Soviet leadership to such an extent that it is unable to order the coordinated launch of its strategic nuclear forces. Minimization of a Soviet attack clearly requires a first strike. Since British targeting is integrated with US targeting policy, it is inevitable that it reflects the new US emphasis on warfighting, and will make full use of the capabilities of the Trident system for accuracy.

A marked feature of British defence policy since the war had been the uniformity of opinion between the Conservative and Labour parties. By the early 1980s this consensus had come to an end. The importance of defence as an issue had been highlighted by the Labour Party's decision at its Annual Party Conference in September 1982 to adopt a policy of unilateral nuclear disarmament while nevertheless remaining within NATO. The decision did not please everyone within the Party. The former Labour Prime Minister, James Callaghan, argued that the policy was 'attractive but wrong' (*The Guardian*, 19 November 1982). In fact, the Party's commitment to unilateral nuclear disarmament was not as clear-cut as it initially seemed. Its Election Manifesto proposed cancelling Trident and refusing permission for the siting of US cruise missiles in Britain, but there was no clear commitment to get rid of the Polaris force, only to include the force 'in the nuclear disarmament negotiations in which Britain must take part'. This essential compromise was to become an embarrassing matter of dispute during the campaign between the unilateralist Party Leader Michael Foot and the Shadow Foreign Secretary Denis Healey who had a strong belief (since radically modified) in the ultimate deterrent effect of nuclear weapons. Just as they appeared to reach an agreement, ex-Labour Prime Minister James Callaghan publicly repudiated the defence policy of his own party. As the Campaign progressed, it became clear that the Labour Party's commitment to unilateral nuclear disarmament was losing it votes even though the electorate as a whole was not in favour of the deployment of either Trident or cruise missiles.

The Conservative victory in the June 1983 General Election appeared to confirm the decision to deploy Trident and Cruise but did not solve the fundamental problems facing Britain's defence policy. Mrs Thatcher's Government had concluded that Britain could continue to carry out the four basic defence tasks which remained following its withdrawal from its world role. These were:

1 the defence of the British Isles;
2 the maintenance of a strategic nuclear deterrent, officially as part of NATO's nuclear forces;
3 the commitment to maintain the British Army of the Rhine and the airforce in support of NATO's central front in West Germany;
4 a naval contribution to the defences of the north-east Atlantic.

However, to these commitments had been added the cost of retaking the Falkland Islands in the 1982 war with Argentina and of continuing to defend them. Defence Secretary John Nott presented a paper to the Cabinet in September 1982 in which he argued that all four of these roles could not be sustained beyond 1986. One option would be to cancel Trident while eking out the life of Polaris until the end of the century (and there are indications that this could be done by careful husbanding of the submarines and acceptance of increased maintenance costs). This alone might not be enough. Another option, proposed by John Silkin, then Labour Party spokesman for defence, was to withdraw from Germany and concentrate on a naval role within NATO. However this would leave Germany as the dominant military power in Western Europe. For historical and other reasons this would cause consternation both to France and the Soviet Union and would be the least favoured option so far as the USA is concerned. Meanwhile, the Conservative Government is not yet ready to make the choice, but is instead prepared to sustain a higher level of military expenditure than any of Britain's major European allies, whether the expenditure is considered in absolute terms, as a proportion of national income, or per head of population.

During 1983 and 1984 the Labour Party's National Executive Committee reaffirmed its commitment to a new defence policy. In July 1984 it proposed the scrapping of Trident, the decommissioning of Polaris, support for a non-nuclear defence policy within NATO, and the removal of US nuclear bases from Britain. Defence spending should be sufficient to provide credible defence, but British and NATO defence policy should seek to reassure the Soviet Union 'that the West is not engaged in an aggressive search for superiority and a relentless arms build-up with which to threaten them'. The Party's immediate goals in respect of NATO were spelt out as, firstly, securing NATO agreement to no-first-use of nuclear weapons; secondly, the creation of a European nuclear free zone; thirdly, withdrawal of all battlefield nuclear weapons from Europe; and, fourthly, the abolition of 'dual-capable' systems. A Labour Government would in any case 'take independent action to remove all battlefield nuclear weapons from British Army and Royal Air Force units in Germany'. Two defence roles would be phased out: Britain's independent nuclear capability, and its 'out of area' role, specifically in the Falklands. It would not, however, withdraw from Germany and would maintain a strong navy, emphasize coastal defence, and favour air-defence squadrons in the RAF. This policy was subsequently endorsed at the Labour Party Conference in October.

The autumn of 1984 saw the Liberal Party Conference vote, against the wishes of its leadership, for the immediate removal of US cruise missiles from Britain. It also called for the immediate cancellation of the Trident programme and for the pursuit of a non-nuclear defence structure under European control. Polaris would be retained but be included in resumed East–West arms-control negotiations. While the vote against cruise showed that the Liberal Party was divided on the issue (611 votes to 556), the effect of the vote was to emphasize the differences between the Liberal Party and its partner, the Social Democratic Party which, while against Trident, continues to adhere to a nuclear defence policy.

Some sections of the Conservative Party, including, according to a *Sunday Times* report (23 September 1984) at least one Minister, have opposed Trident on grounds of cost, and some (for example a pressure group called Tories against Cruise and Trident or TACT) have also opposed the siting of US cruise missiles in Britain, but the Party's leadership has continued to regard both programmes as necessary. Speaking to the Conservative Party Conference in October 1984, Defence Secretary Michael Heseltine stated that Trident was a vital insurance policy although he conceded that the overall cost was high. He also defended the decision to deploy US cruise missiles in Britain. Nevertheless, there is a sizeable lobby among Conservative and other MPs that is moving to the view that a deterrent based on large numbers of British-controlled cruise missiles would be both a cheap and an effective alternative to Trident.

Arguments for the Retention of British Nuclear Weapons

Various arguments have been advanced in favour of an independent British nuclear deterrent. These are most clearly laid out by Jeff McMahan (1981).

The 'Second Centre of Decision'

This argument, originally formulated by Denis Healey, is the one which has been used most frequently by governments to justify an independent British nuclear deterrent. The argument was explicitly stated at length by Francis Pym speaking to the House of Commons on 24 January 1980. He said that, while the USA had consistently made clear its total commitment to help defend Europe, 'the decision to take nuclear action, at any time, would be vastly hard for any President of the United States to take.' While the British Government did not doubt 'the weight and reality of the United States commitment', the Soviet Union 'might be tempted to gamble on United States hesitation' to, in effect, destroy itself on behalf of the defence of Europe. By having its own nuclear deterrent, Britain adds to the credibility of the deterrent by facing the Russians with two rather than only one independent nuclear decision maker within the NATO alliance.

This 'second centre of decision' is said to provide additional uncertainty in the minds of the Soviet leaders in respect of the alliance's commitment to deterrence, since even 'if in some future situation Soviet leaders imagined that the United States might not be prepared to use nuclear weapons, having to take account of the enormous destructive power in European hands would compel them to regard the risks of aggression in Europe as still very grave.' Since 'in practice no other member [of the alliance] is in a position to provide this [second centre]... to give up our distinctive [British] capability now or to let its effectiveness fade away, whether as a deliberate act of unilateral disarmament or simply to save money, would be to abandon not a marginal capability but a central and unique component of our contribution to the alliance.

This would be a particularly strange act at a time when Soviet power relative to the West's is greater than ever before, and still growing' (Cmnd. 8212, 1981, pp. 11–12).

Critics of this argument point out that it rests on the assumption that Britain's independent use of its nuclear deterrent in a situation in which the USA might hang back is credible. However, the inevitable destruction of Britain in a retaliatory Soviet strike would be such that the independent first-use of British nuclear weapons could never be credible. As Lord Carver, the former Defence Chief of Staff put it:

> I have never heard or read a scenario which I would consider to be realistic in which it could be considered right or reasonable for the Prime Minister or Government of this country to order the firing of our independent strategic force at a time when the Americans were not prepared to fire theirs – certainly not before Russian nuclear weapons had landed in this country (Carver, 1979).

Some people argue that Britain needs to retain an independent means of retaliation against the Soviet Union but if Britain is totally destroyed, it could be argued that it would be too late to use nuclear weapons even in retaliation. They would have proved inadequate as a deterrent, and they would not then be weapons of defence; their use would be a form of gratuitous and violent revenge. If Britain were only partially destroyed, their use might provoke a further and more devastating attack. Britain has insufficient weapons and is too small geographically to think in terms of controlled response and intra-war deterrence. A retaliatory strike is not a credible option for Britain under any circumstances.

It is not even clear that the existence of independent British and French nuclear forces weighs very heavily in Soviet calculations. According to Lawrence Freedman (1980, p. 133), the Soviet Union acknowledges the destructive power of the forces, but points out that in any nuclear exchange between the USSR and a second-class nuclear power, the latter would come off worse. The Soviet comment on Pym's 24 January 1980 speech to the House of Commons was that 'people in glass-houses shouldn't throw stones' (Freedman, 1980, p. 133).

It has, moreover, been suggested by Jeff McMahan that the existence of an independent British nuclear deterrent might tempt the USA to hold back its own weapons in the expectation that the British will fire first, thus providing the USA with valuable information about Soviet intentions while minimizing the risk of a Soviet retaliation against them. Given Soviet statements to the effect that any nuclear attack from the West will be met by massive retaliation against all countries harbouring nuclear weapons, this seems unlikely. However, as McMahan points out, it is 'entirely possible that... since [both the US and Britain] would prefer the other to fire first, each might refrain from firing in the hope that the other *will* fire first' (1981, p.24). The resulting confusion and indecision might actually weaken deterrence.

Reliance on the threat to use nuclear weapons independently where the number of such weapons are limited does not seem a credible option. The essential weakness of this posture was recognized by Robert McNamara in his famous speech delivered at Ann Arbor:

> Relatively weak nuclear forces with enemy cities as their targets are not likely to be sufficient to perform even the function of deterrence. If they are small, and perhaps vulnerable on the ground or in the air, or inaccurate, a major antagonist can take a variety of measures to counter them. Indeed, if there was a substantial likelihood of it being used independently, this force would be inviting a 'pre-emptive' first strike against it. In the event of war, the use of such a force against cities of a major nuclear power would be tantamount to suicide, while its employment against significant military targets would have a negligible effect on the outcome of the conflict... In short then, limited nuclear capabilities, operating independently are dangerous, expensive, prone to obsolescence and lacking in credibility as a deterrent (McNamara, 1962).

The 'Trigger'

It has been suggested that, while the second-centre argument is implausible, the independent use of British nuclear weapons might act as a catalyst to draw a reluctant USA into a nuclear war. This adds to the credibility of deterrence because the USSR will ensure that the USA is inevitably drawn into a nuclear war in defence of Western Europe. The idea is that Britain could effectively force the USA into a war it did not want. This could be done in two ways: firstly by openly using British nuclear weapons with the intention of drawing the USA into a nuclear war, while recognizing that this might 'decouple' an angry and alienated USA from the defence of Europe; or, secondly, by deceiving the USSR into believing that the USA and not Britain initiated the attack. While this second course of action would be possible by using Polaris to launch an attack, it would also be possible for the USA to deceive the USSR into believing that the British had initiated an attack.

The trigger argument is important if it is assumed that, in the last resort, the USA will not maintain its nuclear guarantee. The British Government is understandably reluctant to admit to any such doubts in the Alliance. However, Conservative MPs such as Geoffrey Pattie have in the past been less circumspect. In a memorandum submitted to the House of Commons Expenditure Committee in 1978/79, he said 'It is no more than a blinding glimpse of the obvious to say that a guarantee which is no longer automatic is no longer a guarantee and despite the presence of US forces on the ground in Europe in no way can there now be said to be an American nuclear guarantee protecting Western Europe' (Pattie, 1979). In the end, it is inconceivable, given the risks involved, that the USA would act otherwise than in its own interests. If its interests were best preserved by abandoning Europe, then Europe would be abandoned (see p. 120).

The 'Free-Rider'

Here, it is argued that any renunciation of nuclear weapons by Britain would increase the country's dependence on the American nuclear guarantee, and that it would be unfair of Britain to shelter under the US nuclear umbrella without also sharing the risks of its own defence. This position was spelt out in the *Statement on the Defence Estimates 1981* in which it was argued that 'the debate [on British nuclear weapons] should recognise that positions which seek to wash British hands of nuclear affairs, while continuing (as NATO membership implies) to welcome United States nuclear protection through the Alliance, offer neither moral merit nor greater safety' (Cmnd. 8212, 1981, p. 14).

McMahan refers to this argument as the 'free-rider' argument since the claim is that 'in giving up nuclear weapons, Britain would be then getting a "Free-ride" on the back of America's strength' (1981, p. 39). The belief that Britain could shed the risks of nuclear defence would only be credible if it not only gave up its independent nuclear deterrent but also shut down US nuclear bases in Britain and explicitly removed itself from US nuclear protection.

In assessing the validity of the 'free-rider' argument, it is worth noting that of the 15 countries allied to the USA through NATO, only two (Britain and France) have an independent nuclear force. Canada has already opted for unilateral nuclear disarmament, and has almost completed the process (retaining only low-yield and nearly obsolete anti-aircraft nuclear missiles). Iceland has no military forces at all, but allows the US to have bases there. Only 8 of the 15 countries allow the US to station nuclear weapons on their territory; 7 do not. Five do not permit any foreign military bases to be established on their territory (Norway, Denmark, France, Luxembourg and Portugal, although Portugal provides base facilities in the Azores). France and Spain have expelled US nuclear bases. Only 5 European NATO countries agreed to accept US cruise and Pershing II missiles. France, while a member of NATO, withdrew from its military organization in 1966. Britain is the only member of the NATO military alliance, apart from the US, that has its own nuclear weapons. A non-nuclear Britain within NATO would not be unusual.

Moreover, so far as the 'free-rider' argument goes, it should be clear that the USA stays within NATO because it perceives the alliance as serving its interests. Sovereign states do not maintain alliances that are contrary to their interests. Maintenance of the alliance is not dependent on Britain retaining nuclear weapons. The USA has in the past regarded the extension of its 'nuclear umbrella' as a means of preventing nuclear proliferation. Throughout the 1960s the USA stressed the importance of NATO increasing conventional defences in Europe, and opposed the acquisition of independent nuclear forces by other NATO countries. John Nott (1982), however, rejected this view: 'in the last resort, Great Britain must be responsible for her own defence. She should not shuffle that off onto another nuclear power'. The inconsistency of the position is demonstrated by questioning whether or not this means that other countries should reject the US nuclear umbrella and develop their own independent nuclear forces (for example, West Germany) or, alternatively, why Britain accepts US bases over which it can exercise no operational control.

Nevertheless, there would be some strength in the accusation of 'free-riding' if Britain failed to adopt a viable non-nuclear defence policy and at the same time clearly remained dependent on the USA for its protection.

'Nuclear Blackmail'

It is sometimes argued that renunciation of nuclear weapons would leave Britain open to nuclear blackmail by the Soviet Union or some other nuclear power. The threat might be used during wartime (for example to pressurize Britain to withdraw her troops from central Europe), or to obtain compliance with some other policy. Yet the possession of a relatively small independent nuclear force, while it might dissuade other minor nuclear powers from threatening Britain, would not necessarily dissuade the Soviet Union or another major nuclear power from nuclear blackmail, simply because it would not be rational for a minor nuclear power to retaliate against a major one. Much would depend on the evaluation of the benefits to be gained from exercising nuclear blackmail, as against the potential losses if the attempt misfired. Moreover, to argue that one should keep an independent nuclear force to deter nuclear blackmail suggests that other countries should do the same, thus implicitly endorsing proliferation.

The 'European Deterrent'

An underlying fear behind British policy is that the USA might at some time abandon Europe. It is often said that Britain's European allies welcome the British deterrent as a back-up defence which provides some measure of insurance against the failure of the American nuclear guarantee. On the other hand, Lord Carver (1979) stated that 'I have never heard an authorative [sic] military or political figure of any of our allies welcome the fact that we have an independent strategic force,' although he conceded that 'I have indeed heard [our allies] welcome the fact that we are in the nuclear business because, whether they are Americans or European, they believe that the fact that America is not alone in this is of great political value. But it is not necessarily of military value.' The essential military weakness of the 'European deterrent' argument is that the size of the British and French deterrents are such that it could never be rational to use them unilaterally (or in a bilateral Anglo-French attack) against one of the superpowers.

Britain's 'Contribution to NATO'

The British Government stresses that its nuclear weapons are a significant contribution to NATO, and that its renunciation of nuclear weapons would impair the effectiveness of the alliance's defences. NATO, so the argument goes, provides the only realistic way of countering the threat from the Warsaw Pact by deterring aggression. The 1983 *Statement on the Defence Estimates* made it clear that, in the Government's view,

> the British contribution to this collective deterrence, both in terms of our conventional deterrence and in making available bases for United States forces, is of crucial importance to the Alliance. Should we withdraw these forces and close the bases, not only would NATO's ability to deter be substantially weakened, but the cohesion of the Alliance would be put seriously at risk. The result would be to undermine stability in Europe, increase the danger of war and thus jeopardise our own security (Cmnd. 8951, 1983, p. 2).

Yet, in fact, Britain's nuclear contribution to NATO is insignificant when compared with the US forces. It only retains importance in the absence of an American nuclear guarantee and, as has been indicated above, there are grounds for challenging the credibility of an independent nuclear deterrent except perhaps as a means of protecting Britain against nuclear blackmail.

Within the context of NATO there are cogent reasons for arguing that Britain's insistence that it should retain a nuclear capability undermines the security of the alliance by diverting effort away from conventional defence. It does little to enhance the credibility of NATO's deterrent given the strength of the American 'nuclear umbrella' and the weakness of the 'second centre of decision' argument. And it does nothing whatsoever to raise the nuclear 'threshold' by strengthening NATO's conventional defence forces.

The 'Status' Argument

It is sometimes asserted that the possession of nuclear weapons enhances Britain's standing in the world, and underlies her status as a world power. This seems doubtful. Status as a world power depends on economic power and the ability of a country to project its power within the world through alliances and, if necessary, through direct intervention. The depletion of Britain's conventional forces, and its economic decline, have marked its fall from first-power status. The possession of nuclear weapons does nothing to enhance the country's ability to project its power nor is there any evidence that possession of nuclear weapons has enhanced the prestige in which the country is held. Indeed, it may impede its standing as a moral force within the world.

In the past it has been claimed, notably in the 1960 Defence White Paper (Cmnd. 952), that Britain's possession of nuclear weapons 'substantially increases our influence in negotiations' for arms control and disarmament, and in NATO. In fact, the exclusion of Britain from the Strategic Arms Limitation Talks (SALT) and the Strategic Arms Reduction Talks (START) suggests that British influence has been minimal, although it should be said that Britain was anxious to stay clear of SALT in order to exclude forward-based systems from the talks and protect its own independent forces from being bargained away. The British position thus conformed to what Robin Cook (1979) has called the multilateralist paradox: that one must first have a bomb of one's own in order to play a constructive part in banning it. Britain's own record in disarmament and arms-control negotiations suggests that this is not the case.

Conclusions

The history of British defence policy since the war has been characterized by rising costs against a background of economic decline and a withdrawal from a wider world role. By the 1970s Britain's defence policy was decidedly Eurocentric, based on the direct defence of the British Isles, a major land and air contribution to the defence of the European mainland, and the deployment of a major maritime capability in the Eastern Atlantic and the Channel. Within this context, nuclear weapons have been seen as a relatively cheap form of defence.

The dilemma which Britain faces as it enters the mid-1980s is whether or not it can sustain the four basic defence roles which emerged during the 1970s and were clearly identified in Nott's *The*

Way Forward (Cmnd. 8288, 1982) (see p. 89). Three of these roles (and the Government argues that they are the most expensive to maintain) are broadly conventional; only one, the maintenance of Britain's strategic deterrent, is clearly nuclear. It is also the least expensive of the so-called 'mission' programmes (excluding 'other army combat forces'), consuming 2.8 per cent of the defence budget in 1985/86 (see table 4.1).

Although there are strategic arguments for a thorough review of British defence policy, leading to the denuclearization of Britain's armed forces, the cost implications of an enhanced conventional defence policy are considerable. Malcolm Chalmers (1983, p. 65) has estimated that Britain's nuclear programme will consume from 10 to 15 per cent of the total defence budget by the late 1980s when spending on the Trident programme will be at its height. To bring down the cost of Britain's defence to a level equivalent to that of

Table 4.1 The defence budget 1985–6

Total budget £18,060 m.	
Mission programmes	
Nuclear strategic force	2.8
Naval general purpose combat forces	13.9
European theatre ground forces	15.3
Other army combat forces	1.1
Air force general purpose forces	20.5
	53.3
Support programmes	
Reserve and auxiliary formations	2.0
Research and development	12.7
Training	7.1
Repair and storage facilities in the UK	5.1
War and contingency stock	3.0
Other support services	16.5
	46.4

Source: Cmnd. 9430, 1985, p. 34.

its major European allies would require considerable cuts in conventional defence. One solution would be to reduce the objectives of Britain's defence policy by, for example, concentrating on the maritime role or on the land role within Germany. As Chalmers (1983, p. 66) comments:

> By conducting the debate solely in terms of these conflicting commitments, and therefore, services, the military and political establishment has encouraged a paralysis in strategic thinking. For no government, committed to membership of NATO, would find it easy to abandon either the East Atlantic role or the British Army of the Rhine. The problem of finding further cuts in defence spending ... is made to appear insoluble, and the ambitious targets for reducing the defence burden impractical.

Chalmers suggests that the solution lies in adopting an alternative approach, that of non-provocative defence. This option is discussed in chapter 10 (pp. 250–4).

References and Further Reading

Baylis, J. (ed.) (1977) *British Defence Policy in a Changing World*, London: Croom Helm.

Carver, Field Marshal Lord (1979) House of Lords debate, 18 December.

Chalmers, M. (1983) *The Cost of Britain's Defence*, Bradford: University of Bradford School of Peace Studies/London: Housmans.

Churchill, W. (1955) House of Commons debate, 1 March.

Cook, R. (1979) 'Buying a new H-bomb – the easy way', *New Statesman*, 12 January 1983.

Cmnd. 9075 (1954) *Statement on Defence*, London: HMSO.

Cmnd. 9391 (1955) *Statement on Defence*, London: HMSO.

Cmnd. 124 (1957) *Defence: Outline of Future Policy*, London: HMSO.

Cmnd. 363 (1958) *Report on Defence: Britain's Contribution to Peace and Security*, London: HMSO.

Cmnd. 537 (1958) *Agreement for cooperation on the uses of Atomic Energy for Mutual Defence Purposes*, London: HMSO.

Cmnd. 952 (1960) *Report on Defence,* London: HMSO.

Cmnd. 1915 (1962) *The Bahamas Meetings*, London: HMSO.

Cmnd. 2270 (1964) *Statement on Defence*, London: HMSO.

Cmnd. 2592 (1965) *Statement on the Defence Estimates 1965*, London: HMSO.

Cmnd. 2901 (1966) *Statement on the Defence Estimates 1966*, London: HMSO.

Cmnd. 4290 (1970) *Statement on the Defence Estimates 1970*, London: HMSO.
Cmnd. 8212 (1981) *Statement on the Defence Estimates 1981*, London: HMSO.
Cmnd. 8288 (1982) *The Way Forward*, London: HMSO.
Cmnd. 8951 (1983) *Statement on the Defence Estimates 1983*, London: HMSO.
Cmnd. 9430 (1985) *Statement on the Defence Estimates 1985*, London: HMSO.
Crossman, R. (1958) House of Commons debate, 29 February.
Eden, A. (1960) *Full Circle*, London: Cassell.
Freedman, L. (1980) *Britain and nuclear weapons*, London: Macmillan, for the Royal Institute of International Affairs.
Gowing, M. (1974) *Independence and Deterrence. Britain and Atomic Energy 1945–52*, London: Macmillan.
Groom, A. J. R. (1974) *British Thinking about Nuclear Weapons*, London: Frances Pinter.
Joint Statement issued by President Carter and Prime Minister Margaret Thatcher, 18 December 1979.
Macmillan, H. (1955) House of Commons debate, 2 March.
McNamara, R. S. (1962) Address given at the University of Michigan, Ann Arbor, 16 June, cited in Groom, A.J.R. (1977) 'The British Deterrent', in J. Baylis (ed.) (1977) *British Defence Policy in a Changing World*, London: Croom Helm.
McMahan, J. (1981) *British Nuclear Weapons: For and Against*, London: Junction Books.
Mulley, F. (1968) House of Commons debate, 21 March, *Hansard* vol. 946, col. 1315.
Nott, J. (1982) House of Commons debate, 29 March.
Pattie, G. (1979) Memorandum to the Expenditure Committee, *Sixth Report from the Expenditure Committee, Session 1978–9*, p. 129.
Wilson, H. (1966) House of Commons debate, 24 May.

5
Dynamics of the Nuclear Arms Race

There can be little doubt that the arms race has intensified. Both the USA and the Soviet Union feel threatened and are attempting to develop new weapons systems while, on the other hand, engaging in arms-control negotiations. This chapter seeks to explore some of the theories that have been advanced to explain the phenomenon of the arms race.

Game Theory

In its simplest form game theory is applied to situations in which two players are competing against each other. Neither player knows the strategy of the other. Both can make a limited number of moves as a result of which they may benefit (a 'positive pay-off'), lose (a 'negative pay-off') or draw (a 'zero pay-off'). In games of pure conflict one side's gain must always equal the other side's loss (a 'zero-sum game') and game theory teaches that to obtain the best results for himself each player must assess each move in terms of the result that is best for him regardless of what his opponent may do. Games that reflect international relations are not, however, best represented as games of pure conflict since there are possibilities for cooperation as well as conflict, both sides could win or lose simultaneously, and one side's gain does not always equal the other's loss.

One game called Prisoner's Dilemma is often used as a model of the arms race. The prisoner's dilemma arises from a situation in which two people are charged with a crime. They cannot talk to each other. They are told that there is not quite enough evidence to convict them but that if both of them confess they will be convicted. If neither of them confess they will go free. But if one of them confesses and the other stays silent, the first will not only go free but will get a reward, while the other will get an even longer sentence than would be the case if they both confessed. The question applied in the game is, Should a rational prisoner confess or not? Their joint interests are obviously best served if both stay silent but from an individual point of view it makes sense to confess because if the other also confesses at least the sentence is less severe, while if the other stays silent one reaps all the benefits of confession.

Applied to the arms race the game can be displayed on a matrix in which each player (A and B) has two choices, to arm or disarm (see figure 5.1) and in which the relative benefit to each player when combined with the pay-off derived by the other player is

		B's choice	
		Disarm	Arm
A's choice	Disarm	Both disarm A3, B3	Arms advantage to B A1, B4
	Arm	Arms advantage to A A4, B1	Both arm = arms race A2, B2

Figure 5.1 Prisoner's Dilemma applied to the arms race

shown by the outcomes 1 to 4, with the more highly preferred outcome assigned the higher score. The outcomes show that the side which gains an arms advantage will enjoy a substantial lead over its opponent (4 to 1). The fear that the other player is seeking an advantage leads both to reject the benefits of mutual disarmament where their joint score would be greatest (3 + 3 = 6). Each plays safe and arms himself, with they result that the overall outcome is 2 + 2 = 4.

Peter Bennett and Malcolm Dando (1982) have used a theoretical device which they called the hypergame to analyse the arms race in terms of the perception each side has of its own and the other side's motives. First of all they recast the basic model (figure 5.1) to show the game from the point of view of two genuinely

		B's choice: Disarm	B's choice: Arm
A's choice	Disarm	Both disarm A4, B4	Arms advantage to B A1, B3
	Arm	Arms advantage to A A3, B1	Both arm = arms race A2, B2

Figure 5.2 The arms race, played by two genuinely peace-loving players

Source: Bennett and Dando (1982), p. 180.

peace-loving players (figure 5.2). Here, mutual disarmament is seen as the very best option and the option of arming onself while the other disarms is seen as the second-best option. Bennett and Dando then combine the perceptions portrayed in figures 5.1 and 5.2 into their hypergame (figure 5.3) in which each side believes that its own ranking of options springs from genuinely peace-loving attitudes (i.e. conforms with the matrix shown in figure 5.2) while that of its opponent accords with the ranking shown in the prisoner's dilemma model of the arms race (figure 5.1) in which aggression is valued more highly than cooperation.

The problem with game theory explanations of the arms race is that they are over-simple. Bennett and Dando accept this but

		Game seen by A		*Game seen by B*			
		B's choice: Disarm	B's choice: Arm	B's choice: Disarm	B's choice: Arm		
A's choice	Disarm	A4, B3	A1, B4	A3, B4	A1, B3	Disarm	A's choice
	Arm	A3, B1	A2, B2	A4, B1	A2, B2	Arm	

Figure 5.3 Bennett and Dando's arms race hypergame

Source: Bennett and Dando (1982), p. 181.

point out that it is precisely this kind of incredibly simple-minded game-theory model of the world that informs defence policy. The approach nevertheless highlights some of the assumptions underlying the arms race, notably the fear that one's opponent will try to gain an advantage by out-spending one in arms research, development and procurement.

Action–Reaction

The idea of an arms race encapsulates the concept of competition between nations and the view that each is trying to gain superiority in military strength. Until the First World War, national power was equated with military manpower and the size of the population from which it was drawn, and with the general effectiveness and direction of a country's communication and transportation systems (notably its railways). Only in the late nineteenth- and early twentieth-century naval arms race was the application of technology seen as providing a quantitative and qualitative advantage.

During and after the First World War the application of technology to military ends in the form of the motor vehicle, the tank, the aeroplane and, ultimately, the missile led to increased competition first between the European powers and subsequently between the superpowers. Since the Second World War the USA has consistently led in the technological arms race. On those few occasions when the Soviet Union has succeeded in achieving a 'first' (for example, in the flight testing of ICBMs and the launching of a man-made satellite) its lead has rapidly been eclipsed by the USA. In spite of the periodic 'scares' when the Soviet Union is depicted as being ahead in the arms race, most commentators agree that it lags behind the USA in military technology.

Because the Soviet Union generally lags behind the West, the action–reaction model of the arms race does much to explain the general direction taken by the Soviet armaments industry. Mikhail Agursky, a Soviet *émigré* who worked in the Soviet defence industry, writes, with Hannes Adomeit, that 'the pace and direction of military technological innovation' has been set 'primarily in the Western countries, notably the United States' (Agursky and Adomeit, 1979). David Holloway (1984, p. 148) has similarly suggested that 'Soviet military R & D in the post-war period can be seen as the effort of a basically non-innovative system to cope with revolutionary technological change, which has been generated primarily by the Soviet Union's potential enemies'.

The Military-Industrial State

The Armaments Industry in the West

At any time the military procurement budget of most Western countries is dominated by a few major 'weapons systems', a concept that embraces not only the weapon's platform (ship, tank or aircraft), the weapon itself (gun, missile or torpedo) and the means of command and communication, but also the 'entire supporting cast – scientists to invent the weapons, workers to build them, soldiers to use them, and technicians to repair them' (Kaldor, 1982, p. 12). The design and production of such systems is undertaken by a handful of companies known as 'prime contractors' who specialize in particular weapon systems. Boeing, General Dynamics and Rockwells are bomber enterprises. Grumman and Vought make fighters for the US Navy. McDonnell Douglas and General Motors are prime contractors for battle tanks. Westland (in Britain), Sikorsky, Bell and Boeing Vertol (in the USA) and Aerospatiale (in France) make helicopters. Electric Boat has made submarines since the 1890s. The design and production of particular subsystems (e.g. gun, missile, engine, electronics) is often subcontracted to other companies, who are to a greater or lesser extent dependent on the prime contractors. Such companies include Rolls Royce, United Technologies, and General Electric (engines), and Texas Instruments, Raytheon, Westinghouse, Ferranti and Marconi-Elliot (electronics). Many of the companies depend for their survival on defence contracts. From the companies' point of view it is vital to maintain the 'follow-on' system in which, as soon as design work on one weapon system is finished and goes into production, the company begins to design the 'follow-on' which should eventually replace it.

From the government's point of view the maintenance of a viable defence industry is equated with national security. Companies are helped by the provision of working capital for projects (including at times the provision of the plant in the form of so-called GOCO or government-owned contractor-operated plants). Governments may also step in to help prime contractors who are in difficulty, either through subsidies (as happened in the case of Lockheed when it was brought to the verge of bankruptcy by the cost overrun on the B-1 bomber) or in the form of new defence projects.

In his farewell speech to the American people, broadcast on 17 January 1961, President Eisenhower (1961, p. 421) warned of the dangers of what he called the military-industrial complex:

> we can no longer risk emergency improvisation of national defence; we have been compelled to create a permanent armaments industry of vast proportions...
>
> This conjunction of an immense military establishment and a large arms industry is now in the American experience. The total influence – economic, political, even spiritual – is felt in every city, every statehouse, every office of the federal government. We recognise the imperative need for this development; yet we must not fail to comprehend its grave implications. Our toil, resources, and livelihood are all involved; so is the very structure of our society.
>
> In the councils of government we must guard against the acquisition of unwarranted influence, whether sought or unsought, by the military-industrial complex. The potential for the disastrous rise of misplaced power exists and will persist.

The pervasiveness of the military-industrial complex and its influence on other spheres of life led American analysts to coin the phrase 'the Iron Triangle' to describe the network of interests which encompass Congress, the Pentagon and the arms manufacturers. A recent study edited by Gwyn Prins (1982, p. 134) has referred to 'the Steel Triangle', a three-pointed 'military-academic-industrial complex' which, while falling far short of a conspiracy, has a momentum of its own. Indeed, one American researcher, Alvin Sunseri (1977, p. 158) believes the network of interests is so pervasive that he has spoken of the 'military-industrial-technological-labour-academic-managerial-political complex' (MITLAMP).

The role of scientists and technologists in the military-industrial complex

Individual scientists such as Lord Zuckerman, science adviser to a succession of British governments in the 1960s, and Jerome B.

Wiesner, President Kennedy's science adviser, have drawn attention to the link between the academic scientific community and the development of weapons. Zuckerman (1982, p. 103) has expressed the view

> derived from many years of experience, that the basic reason for the irrationality of the whole process [of military research, development and procurement] is the fact that ideas for a new weapon system derive in the first place, not from the military, but from different groups of scientists and technologists who are concerned to replace or improve old weapons systems... At base, the momentum of the arms race is undoubtedly fuelled by the technicians in government laboratories and in the industries which produce the armaments.

Thus scientists and technicians anticipate needs long before they are recognized by the military, and produce new devices 'when it is barely possible to perceive their relation to the military operations... then in progress'. Zuckerman argues that it is the nuclear scientists and technicians who start the process of formulating 'so-called military need' and who have been responsible for equating and confusing nuclear weapons with military strength. Indeed, before any decision had been taken by the British Government to replace Polaris with Trident, and long before the need for any such decision had been put to Ministers, British nuclear scientists had begun designing and testing a warhead for a MIRVed Trident missile (*The Sunday Times*, 10 August 1980; *The Guardian*, 6 March 1981).

The role of the universities

In the USA, the volume of Pentagon-funded research in the universities is considerable. In 1980, 203 higher education institutions were contracted by the Department of Defense to a total value of $652 million. To this extent, the political decision to fund military-related research and development precedes and initiates the actual process of research and development. Only at the latter point does 'scientific initiative' take over and deliver more than is strictly required. Policy then adapts to and may even overtake this over-achievement, so setting the process off once more.

The importance of military-funded university research had been widely admitted and dates back many years. Robert C. Seamans

Jr, then Air Force Secretary and a former academic at Massachusetts Institute of Technology and deputy administrator of NASA, said in 1969 that 'we cannot provide the necessary weapons for defense without the help of university research laboratories' (*Philadelphia Inquirer*, 16 May 1969). John Hannah, President of Michigan State University and a former Assistant Secretary of Defense, insisted in 1961 that 'our colleges and universities must be regarded as bastions of our defense, as essential to the preservation of our country and our way of life as supersonic bombers, nuclear-powered submarines and intercontinental ballistic missiles' (cited in S. Lens, 1971, p. 127). The US Department of Defense also helps fund a number of other non-university research organizations and policy review bureaux such as the Rand Corporation.

The role of industry

The arms industry naturally plays an important role in the process. It is far from being a passive supplier of military equipment. David Sims (1970, p. 248) quotes Pete Schenck of the Raytheon Corporation as saying that

> The day is past when the military requirement for a major weapons system is set up by the military and passed on to industry to build the hardware. Today it is more likely that the military requirement is the result of joint participation of military and industrial personnel, and it is not unusual for industry's contribution to be a key factor. Indeed, there are highly placed military men who sincerely feel that industry currently is setting the pace in the research and development of new weapons systems.

Since they are working at the forefront of the technology, such individuals, by stressing the likely technological capabilities of the future, are also able to raise fears about the future military capabilities of the Soviet Union. This process of 'threat inflation' may be linked to incomplete and inflated intelligence reports.

'Threat inflation' assures future business for the armaments firms and can be used by politicians to justify arms procurement programmes. The techniques used are frequently no better than crude (but often effective) advertising, and are exemplified by full-page advertisements such as those placed by the Bath Iron

Works (a US producer of naval vessels) showing Soviet ships being deployed in the Atlantic and warning of the danger of the ocean becoming a 'Red Sea'; and of the British Government's film, *The Peace Game*, pointing to the 'threat' posed by Russian tanks and Soviet SS-20 missiles, and justifying NATO's modernization programme. The aim of such advertisements and propaganda is to influence politicians and the general public and gain their acceptance of the arms build-up.

The links between industry, government and the military

The interchange of military and industrial personnel establishes a network of relationships which binds the military-industrial complex together. Thus, for example, Charles E. Wilson, Secretary of Defense during the 1953–7 Eisenhower Administration, came to the post from General Motors; Robert McNamara left his post as President of the Ford Motor Company to become Secretary of Defense in the Kennedy Administration; and Alexander Haig, former four-star general and Supreme Allied Commander of NATO forces, was President of United Technologies, makers of helicopters and jet engines, until he became President Reagan's first Secretary of State.

In 1973, 3,233 former high-ranking military officers (defined as retired colonels or Navy captains and above) were employed by 100 companies which, between them, accounted for two-thirds of US Department of Defense work. In the same year, 275 former defence industry managers held ranks in the top three levels (GS13 or higher) of government service in the US Defense Department (Prins, 1982, p. 154). A former Assistant Secretary of Defense, J. Ronald Fox (1974, p. 461) has commented that

> The availability of jobs in industry can have a subtle, but debilitating effect on an officer's performance during his tour of duty in a procurement management assignment. If he takes too strong a hand in controlling contractor activity, he might be damaging his opportunity for a second career following retirement. Positions are offered to officers who have demonstrated their appreciation for industry's particular problems and commitments.

As a result, 'If a retired general representing a client goes in to see an old classmate still on active duty, he will get a very attentive hearing. The officers on active duty are also thinking ahead. Fighting the system gets one blackballed and future employment prospects are bleak. In this way the industry has come to completely control DOD [Department of Defense] even more than its political appointees' (T. S. Amlie, n.d. in Fallows, 1981, p. 66). This is not to say that individuals are not impartial, or use their office for private gain, but there is a potential conflict of interest.

Gordon Adams' (1981) study of the widespread and growing political influence of the leading defense contractors in the USA (Boeing, General Dynamics, Grumman, McDonnell Douglas, Northrop, Rockwell International and United Technologies) shows that between 1970 and 1979 they received over 25 per cent of Department of Defense awards (over $100 billion worth of contracts), together with a further $11.4 billions from NASA. He showed that five of these companies over a two-year period spent $16.8 millions on their Washington offices. All the companies mentioned had registered lobbyists and established Political Action Committees between April 1976 and February 1978. By mid-1980 these committees had between them spent $2 million, including $1.26 million in contributions to Federal campaigns. The Political Action Committees also contributed to the expenses of leading politicians on the Senate and House Armed Services Committees, and the Senate and House Defense Appropriations Subcommittees (Adams, 1981, pp. 34–5, 115–16, 137).

The movement of personnel between the Department of Defense and the companies referred to above provides the latter with access to the defense policy process. All the companies play a key role in defining future weapon systems and all have representations on key Department of Defense and NASA advisory committees. A considerable amount of the companies' Research and Development programmes is funded through the Department of Defense Independent Research and Development, and Bids and Proposals programmes. Adams estimates that 78 per cent of Grumman's research and development in the period 1973–8 was reimbursed by the Department of Defense. Other companies had a lower proportion of this cost reimbursed: United Technologies, at the other end of the scale, had only 14.5 per cent reimbursed. Department of Defense Independent Research and Development funds helped Pratt & Whitney develop the JT 9D engine (pp. 79, 97–8, 100–1, 167–71). In all Pratt & Whitney received $87 million for this purpose, yet by 1973, while 1,301 of JT 9D engines had been sold to commercial companies only 3 were sold to the Department of Defense (Daniels, 1980, p. 49).

The companies also try to mobilize opinion on their own behalf. Rockwell spent some $1.35 million between 1975 and 1976 to rally support for the B-1 bomber programme. They also try to affect the climate of opinion among decision makers through trade associations such as the Aerospace Industries Association and the American Defense Preparedness Association, with a view to increasing the level of expenditure on arms. Advertising in trade journals and lobbying at the grassroots level also play their part in the process. The overall result, in Adams' view, creates 'an "iron triangle" on defense policy and procurement that excludes outsiders and alternative perspectives' (Adams, 1981, pp. 156–8, 186–8, 192, 207).

The Soviet Defence Industry

The basic structure of the Soviet defence sector was laid down by the 1940s, and has endured since then with remarkable stability. In contrast to the West, research institutes, design bureaux, and production plants are organized as separate entities. The research institutes are attached to the central defence industry ministries, of which there are believed to be nine.

The design bureaux may be attached to production plants or to research institutes, or be relatively independent. Their budgets are relatively stable, so that the problems faced by Western companies which have to compete for contracts and are subject to market forces, are unknown. Soviet research and development institutes are thus able to pursue their specializations and maintain several different product lines more easily than is the case in the USA. As a result, the number of new weapons systems entering Soviet service has been maintained. Competition is institutionalized, in the sense that successful design bureaux are those which gain production contracts and follow-on assignments and tend also to enjoy a greater stability of output and employment than their counterparts in the West. Excess capacity, where it exists, is a consequence of centralized planning. All the plants produce civilian products which serve as a buffer between defence contracts and can ease the conversion of a particular plant from military to civilian use and vice versa. Each defence industry tends to keep as much of the manufacturing process under its own control as possible, in order to ensure supplies. As a result, the subcontractors are also protected from market forces. The insistence on common designs and design simplicity also helps to maintain supplies and minimize disruption arising from design changes.

The overall work of the military industrial hierarchies is coordinated by the military-industrial commission which is, in turn, coordinated with the military through the Defence Council. The Ministry of Defence issues requirements, requests design proposals, concludes development contracts with design bureaux, supervizes development, conducts prototype trials, places orders for production and assimilates new equipment. Each service has armaments directorates that deal directly with the design bureaux and production plants. They also have research establishments which help the services decide on their operational requirements, monitor scientific and technological developments, test prototypes and advise on new equipment. In general, a decision in principle to produce a new system is taken at an early stage, at the same time as the design is adopted. If state trials are successful, the chances that the new system will be produced are high, but the decision to produce a series does not always follow. Thus, although the first Soviet ICBM, the SS-6, passed its state trials, only four missiles were ever produced.

The centralized nature of planning in the Soviet Union is well suited to the achievement of specified objectives, including implementation of radical design and product changes where this is felt to be necessary. On the other hand, the system encourages waste, over-investment, duplication and redundancy, and discourages innovation and technical progress because the changes necessary to bring about such progress may disrupt existing production and the achievement of quantitative targets.

The Military-Industrial Complex in the Soviet Union

Although the Soviet Union possesses a large defence industry and powerful and well-resourced armed forces, Holloway (1984, p. 109) points out that final control rests with the Politburo which makes the major decisions about resource allocations. He argues that while close ties exist between the military and the defence industry 'it is still the vertical relationships, culminating in the

Party leadership, that predominate ... the build-up of military power since the late 1950s must be seen as the product of conscious political choice, and not only as the result of pressure from a military-industrial complex' (p. 159). To this extent Holloway suggests that there is no evidence of a military-industrial complex comparable with that found in the West (p. 159).

On the other hand, Khrushchev complained in his memoirs (1971, p. 519) that 'the leaders of the armed forces can be very persistent in claiming their share when it comes to allocate funds'. Garrison and Shivpuri (1983, p. 299) point out that both President Brezhnev and Defence Minister Ustinov gained their formative experience in the defence-industrial sector and they argue that 'the defence sector in the Soviet Union is the most powerful pressure group just as it is in the US or NATO' (p. 299). There is, they say, fierce competition between the military services, the various design bureaux and the manufacturing firms and one cannot assume that the pressures found in the West do not also occur in the Soviet Union (p. 299).

Bankrupting the Soviet Union

The drive for military superiority and the increase in US defence expenditure pose a direct challenge to the Soviet Union to which the Soviet leaders have responded. However, while arms expenditure in the USA tends to lead to an upsurge in the US economy, and has indeed been used by governments as an economic regulator to promote growth, in the Soviet Union it acts as a depressor, diverts resources away from the civilian sector, and so encourages consumer discontent.

Reliable data on the level of military expenditure by the Soviet Union and its allies is not available. Nevertheless, Western governments have frequently justified their own arms build-up by pointing to the level of military expenditure in the Soviet Union. US Secretary of Defense Weinberger (1982a) argued that the global military balance 'has shifted against [the US] because the Soviet Union has out-invested and out-produced us for at least a decade'. Such comparisons are of dubious validity. Dollar comparisons of US–Soviet military expenditure generally fail to take account of the different cost-structures applying in the Soviet Union (for example, the average cost of a serviceman is much lower in the Soviet Union than it is in the US). They tend to ignore differences in the productivity and utilization of capital equipment, in both of which the Soviet Union is known to be less efficient than the USA. They also ignore differences in force structure (for example, in the ratio of front-line to non-fighting troops) and in the relative effectiveness of the equipment available to each side (US equipment is technologically more advanced than that of Soviet forces). Account also needs to be taken of the external threats which the forces are designed to counter; the Soviet Union maintains considerable forces in the Far East along its border with China.

There is little doubt that military spending is a heavier burden on the Soviet economy than on the US one. The USSR's national product is substantially smaller and it has been estimated that military production accounts for 25 per cent of industrial production. Moreover, while the US, in common with most Western economies, has a large pool of unemployed labour, the USSR is experiencing a labour shortage. Overall the opportunity cost of military spending – that is the amount of civil output foregone – is much higher in an economy which has full employment than in one with high unemployment. The economic growth rate of the Soviet Union has been slowing down, yet there is no sign that the upward trend in military spending is following suit. The burden of military expenditure therefore appears to be increasing in the Soviet Union.

The Reagan administration has indicated that its increased defence spending has, as one of its objectives, the intention of forcing the USSR into a military-economic competition which it cannot win, and of disrupting the Soviet economy to such an extent that internal social unrest results. The Pentagon's five-year defence plan for 1984–8 ("Fiscal 1984–1988 Defence Guidance") argued that American arms procurement should favour weapons 'that are difficult for the Soviets to counter, impose disproportionate costs, open up new areas of major military competition, and render the accumulated Soviet equipment stocks obsolete' (Halloran, 1982).

The Geopolitical Struggle between the Superpowers

Since 1945 the dominant influence in world politics has been the globalized and bipolar rivalry between the USA and the Soviet Union. This conflict is given added importance because it is *systemic*. 'It is not just one between rival states ... It is a conflict in which aspects of this great power rivalry are grafted onto rivalry between two social systems that remain, with all necessary qualifications, in continuing conflict' (Halliday, 1983, pp. 31–2). Not only are these two systems organized on fundamentally different principles in relation to ownership of the means of production but each 'stake an ideological claim to be world systems, ideal societies which others should aspire to follow' (Halliday, 1983, p. 33).

The Utility of Nuclear Weapons

One of the hallmarks of the geopolitical and ideological struggle between the USA and the Soviet Union has been the arms race. There are those who believe that if the arms race could be stopped and reversed, international tensions would be reduced. Yet the arms race is meaningless without the conflict that fuels it. While it contributes to insecurity and tension, any explanations that concentrate solely on the arms race 'obscure the very real political and social issues around which international affairs have revolved and which have themselves to a great extent determined the course of that arms race' (Halliday, 1983, p. 31). The attempt by the Soviet Union to challenge the global hegemony of the USA through the development of its naval power and the achievement of strategic parity is seen by the USA as an attempt to break its power and establish an alternative Soviet hegemony. Short of war there may be no means of checking the rival except by the mobilization of stronger military power. As Michael Howard (1981, p. 21) suggests 'An arms race then becomes almost a necessary surrogate for war, a test of national will and strength; and arms control becomes possible only when the underlying power balance has been mutually agreed.'

Howard has argued that prior to the seventeenth century warfare consisted of predatory raids against which the defenders could rarely offer protection. With the development of the nation state with its power to defend frontiers and deploy standing armies, such direct coercion was no longer possible. The art of military strategy, which Howard defines as defensive or offensive organized coercion involving the use or threatened use of force to compel an adversary to abandon his preferred course of action and conform to one's will (Howard, 1973, p. 85), developed in Europe to meet the situation in which 'in any conflict the antagonists were in complete control of their national resources and ... commanded a total consensus of national will' (Howard, 1973, p. 88).

At first, he argues, nuclear weapons could be seen as 'an extension of a long trend in military technology' (p. 91). On at least 19 occasions between 1946 and 1973 the USA deployed nuclear weapons in situations of international tension. So long as the USA enjoyed nuclear superiority it could confront the communist powers with 'the inescapable alternatives of unacceptable destruction or compliance with one's will' (p. 91). Once the Soviet Union also had nuclear weapons and an assured means of delivering them 'the strategic situation ... reverted to that of the sixteenth century and earlier epochs, when princes could not prevent their rivals from ravaging their territories but only deter them, by taking hostages or retaliating in kind. Governments again found themselves unable to protect their populations' (pp. 91–2). Not only did nuclear weapons appear unusable but the theory of deterrence appeared to limit the value of conventional forces as a means of coercing or threatening a nuclear armed opponent. The Soviet attainment of strategic parity in the late 1960s and early 1970s thus posed significant problems for the USA at a time when the US was also embroiled in Vietnam. Henry Kissinger (1982, p. 1176) lamented that 'until the early Seventies, in fact, the worst-case scenario analysis of the Soviets was bound to be a significant restraint on adventurism. Therefore, our loss of strategic superiority was a strategic revolution even if the Soviets did not achieve a superiority of their own. For that, to some extent, freed the Soviet capacity for regional intervention'.

Recent technological advances appear to have given strategy a new lease of life. The development of accurate counter-force weapons and the expectation that one day it will be possible to defeat an attack by ballistic missiles through the defence of what the 'ad-men' of the military-industrial complex have dubbed 'The High Frontier' apparently offers the hope that strategy – the art of coercing one's opponent – can be restored to its former position. The threat posed by war-winning nuclear weapons will also enable conventional forces to be used under a nuclear umbrella.

The utility of nuclear weapons is thus said to lie in their enabling function. Eugene Rostow, President Reagan's first Arms Control and Disarmament Agency Director, said in an interview: 'Lots of people don't realize that the purpose of nuclear weapons is not [for them] to be used, but to allow for the use of conventional weapons ... Nuclear arms are a political weapon' (Rostow, 1984). Their importance in the eyes of the first Reagan Administration was further underlined by Richard Perle, Assistant Secretary of Defense for International Security Policy, who argued that

> It is not that I am worried about the Soviets attacking the United States with nuclear weapons confident that they will win that nuclear war. It is that I worry about an American President feeling that he cannot afford to take action in a crisis because the Soviet nuclear forces are such that, if escalation took place, they are better poised than we are to move up the escalation ladder (cited in Scheer, 1982, p. 13).

This explains both the American drive to maintain their superiority and the fear of loss of nuclear superiority that impels the arms race.

Horizontal Escalation

Howard (1976, pp. 36–7) reminds us that, in general, states have their origins in war and that it is war that brings about their decline and eclipse. In the aftermath of the Second World War it became common to see the post-war world 'as one divided into "spheres of influence" with appropriate military power to define and maintain them' (pp. 41–2). The world came to be visualized in terms of possible armed conflict between the superpowers. As a result the USA

> so conducted its policy as to maximize its military effectiveness in the event of such a conflict ... It wooed and armed allies, attempted to intimidate neutrals and set itself the task of building up and maintaining a nuclear strike capability which would enable it to retaliate massively, at times and places of its choosing, to Soviet aggression anywhere in the world (p. 42).

In the 1950s the USA, in its capacity as the world's only *global* superpower, attempted to contain the Soviet Union through a series of military alliances. Intervention in Third World crises led to the Vietnam war. In the aftermath of that war the Nixon Administration could not afford politically to intervene directly in support of its interests and the Administration began to rely on secondary Third World powers to police the world and support US interests. The Iranian revolution shattered this concept and convinced the USA that it could no longer depend solely on others. After 1980, the Reagan Administration began to evolve a new global military strategy. Its aims were spelt out by the US Secretary of Defense, Caspar Weinberger (1982b, p. III-91):

> Our long-term goal is to be able to meet the demands of a world-wide war, including concurrent reinforcement of Europe, deployment to Southwest Asia [i.e. the Persian Gulf region] and the Pacific, and support for other areas ...
>
> Given the Soviets' capability to launch simultaneous attacks in [Southwest Asia], NATO, and the Pacific, our long-range goal is to be capable of defending all theaters simultaneously

Within this context Weinberger's strategy assumes that the USA

> might choose not to restrict ourselves to meeting aggression on its own immediate front. We might decide to stretch our capabilities, to engage the enemy in many places, or to concentrate our forces and military assets in a few of the most critical arenas ...
>
> A wartime strategy that confronts the enemy, were he to attack, with the risk of our counter-offensive against his vulnerable points strengthens deterrence ... Our counter-offensive should be directed at places where we can affect the outcome of the war. If it is to offset the enemy's attack, it should be launched against territory or assets that are of an importance to him comparable to the ones he is attacking (Weinberger, 1982b, p. I-16).

Since the late 1970s American strategists have assumed that the next war will begin in the Persian Gulf. Robert Tucker (1980) has argued that 'the center of gravity of American interests in the world today is not to be found in Europe but in the Persian Gulf'. He points out that despite the Rapid Deployment Force and its bases in the region, the US could not defend the Gulf against a determined Soviet attack without recourse to nuclear weapons. Weinberger's solution to this problem is to threaten to start a war elsewhere in response to a Soviet attack – a strategy known as 'horizontal escalation'.

Weinberger (1982b, p. I-14) specifically indicates that 'For the region of the Persian Gulf, in particular, our strategy is based on the concept that the prospect of combat with the US and other friendly forces, coupled with the prospect that we might carry the war to other arenas, is the most effective deterrent to Soviet aggression.' In this connection he has pointed to Soviet vulnerabilities in Eastern Europe:

> Some important Soviet vulnerabilities have to do with the fact that the Soviet empire, unlike our alliance, is not a voluntary association of democratic nations. Thirty-seven years after free elections were promised at Yalta, the imposition of martial law in Poland makes clear how such elections would turn out if they were permitted. Our plans for counteroffensive in war can take account of such vulnerabilities (Weinberger, 1982b, p. I-16).

Duncan Campbell (1982) drew attention to certain highly classified contingency plans in which it was envisaged that the US might launch a 'pre-emptive attack' and an invasion to 'liberate' East Germany and Czechoslovakia, with or without the support of the NATO allies. Such plans give a new significance to the presence of US missiles and forces in northern and central Europe, particularly since these forces remain outside of the control of the host countries.

Ostensibly the theatre nuclear weapons reflect the USA's commitment to defend Europe and, through the NATO doctrines of flexible response and escalation, ensure that the US strategic nuclear umbrella continues to be extended over Europe. There are many Americans who do not believe that the idea of a US nuclear umbrella is rational. Henry Kissinger, speaking at a conference on the future of NATO held in Brussels in September 1979, argued 'that our European allies should not keep asking us to multiply strategic assurances that we cannot possibly mean, or if we do mean, we should not want to execute, because if we execute, we risk the destruction of civilization' (cited in Johnstone, 1984, pp. 180–1). At the same conference the neo-conservative Irving Kristol said: 'Under no conceivable circumstances – I repeat *no* conceivable circumstances – will an American government respond to such Soviet aggression in Western Europe by initiating a strategic nuclear exchange. *That* deterrence has ceased to exist, whatever some military men or politicians may officially say. The function of America's strategic nuclear weaponry today is to deter a Soviet strike against the United States itself – that and nothing but that' (cited in Johnstone, 1984, p. 182).

The function of US forward-based systems is thus to strengthen US capability to intervene in the Third World by holding the Soviet Union in check through fear of US initiated 'horizontal escalation', and not to defend Western Europe. US policy is to encourage Europe to defend itself by building up its own forces and relying less on American forces. In particular, there has been a marked change in US attitudes towards its allies' nuclear forces. Whereas McNamara dismissed 'relatively weak nuclear forces' as 'not likely to be sufficient to perform even the function of deterrence' (McNamara, 1962), both the Carter and Reagan Administrations have supported the British acquisition of Trident and the maintenance of the UK independent deterrent (see for example the December 1979 Communiqué issued followed Mrs Thatcher's meeting with President Carter, Joint Statement 1979, and President Reagan's letter of 11 March 1982 to Mrs Thatcher, Cmnd. 8517, p. 4). More generally, there is some evidence that the Pentagon now believes that nuclear proliferation is out of control and that in the circumstances it is no longer worth supporting the Non-Proliferation Treaty (Johnstone, 1984, p. 205). In these circumstances the US must ensure that it keeps ahead in the technology.

Nuclear weapons, it is argued, may deter non-nuclear or weak nuclear powers from attacking strong nuclear powers while enabling the latter to intervene militarily in non-nuclear areas. The French Admiral, Antoine Sanguinetti (1981), has drawn attention to this kind of argument:

> One of the postulates of the existing world political system is that the nuclear nations whose territory is now kept safe from any armed attack by the balance of terror are thus enabled to practise an indirect strategy in the rest of the world, going as far as armed action without risk to themselves. Unfortunately, these capacities for indirect strategy in the nuclear age have let loose all sorts of greed in a new reign of force. In practice, nuclear nations help themselves to the right to intervene as they wish by armed might in the affairs of others.

European involvement in the Third World is positively encouraged by the USA which 'sees Third World conflict affecting US interests as much more likely to occur than overt Warsaw Pact aggression against NATO' (Komer, 1984). In part this has already occurred. The US cruise missiles stationed at Comiso in Sicily have nothing 'to do with the defence of Western Europe from Warsaw Pact aggression. Instead, it clearly has everything to do with the establishment of a *de facto* protectorate over the Middle East and northern Africa, and eventually the Balkans as well' (Johnstone, 1984, pp. 145–6). Johnstone also points to US encouragement of European participation in the multinational peacekeeping force in the Lebanon; the shift of Italy's military forces from the north-east of the country to the south and the plans to reorganize the services and train them for rapid intervention; the secret use by the USA of its satellite technology to support British forces in the Falklands War, thus ensuring British success and accomplishing 'the political task of involving British forces in precisely the sort of global policing on which the United States insisted so strongly'; and the attempts being made to get Japan to increase its military spending.

The basis of continued US supremacy rests on its dominant lead in high technology. Johnstone draws attention to an article by Richard Burt, Reagan's director of politico-military affairs at the State Department, which suggests that satellite-guided cruise missiles could be used in Third World conflicts, particularly in theatre warfare where they could deliver nuclear or chemical warheads with high accuracy (Burt, 1981). As Johnstone (1984, p. 202) comments, 'satellite-guided cruise missiles thus seem to be recommended as the ideal weapons system for the high-tech elite countries to use in wars for control of the Gulf oil resources'.

Johnstone argues that the USA needs Europe and Japan to help it maintain its global role. Politically it has raised the spectre of a Europe, abandoned by the USA and unwilling to defend itself, falling prey to Soviet expansionism. Forced to invest in armaments, the Europeans then help to relieve the USA of some of the costs arising from its global military strategy. Politcally, the U.S. encourages its allies to support its strategy. Militarily, Europe provides a base from which US missiles can threaten the Middle East and North Africa, and can pose the threat of horizontal escalation to restrain Soviet action. Financially the USA has, through high US interest rates and the steep rise in the exchange price of the dollar, effectively found a way of forcing its allies to finance its armaments programme.

Conclusions

No single factor explains the current arms race. To some extent it is fuelled by fear although this is not the most important factor; much of the fear derives from threat inflation. To some extent the arms race is the child of the military-industrial complex. There are clearly powerful institutional forces at work which sustain it. To some degree the Soviet Union is reacting to the USA. Chiefly, the arms race is justified by and sustained by the geopolitical and ideological struggle between the USA and the Soviet Union, and derives its importance from the USA's determination to dominate and control the Third World and sustain its globel hegemony. Within this context nuclear weapons are seen by the USA as being a means of threatening the Soviet Union and thus preventing her from challenging US hegemony, and also, increasingly, as of potential use in localized conflicts in the Third World.

References and Further Reading

Adams, G. (1981) *The Iron Triangle. The Politics of Defense Contracting*, New York: Council on Economic Priorities.

Agursky, M. and Adomeit, H. (1979) 'The Soviet Military-Industrial Complex', *Survey: A Journal of East and West Studies*, 24, 2.

Alexander, A. (1973) *Weapon Acquisition in the Soviet Union, United States and France*, Santa Monica: Rand Corporation, Document P-49-89.

Amlie, T. S. (n.d.) 'Defense Acquisitions – Some Observations and Suggestions'. Unpublished paper cited in J. Fallows (1981) *National Defense*, New York: Random House.

Bennett, P. G. and Dando, M.R. (1982) 'The Nuclear Arms Race: Some Reflections on Routes to a Safer World', in B. Newman and M. Dando (eds) (1982) *Nuclear Deterrence: Implications and Policy Options for the 1980s*, Tunbridge Wells: Castle House Publications.

Burt, R. (1981) 'Local Conflicts in the Third World', in R. K. Betts (ed.) (1981) *Cruise Missiles: Technology, Strategy, Politics*, Washington DC: The Brookings Institute.

Campbell, D. (1982) 'America's Base Motives', *New Statesman*, 17 December.

Cmnd. 8517 (1982) *The British Strategic Nuclear Force, Text of Letters*, London: HMSO.

Daniels, M. (1980) *Jobs, Security and Arms in Connecticut: A Study of the impact of military spending on the State*, Voluntown, Connecticut: American Friends Service Committee.

Eisenhower, D. D. (1961) 'Farewell radio and television address to the American people, 17 January 1961' in *Public Papers of the Presidents of the United States, Dwight D. Eisenhower 1960–61*, Washington DC.

Fallows, J. (1981) *National Defense*, New York: Random House.

Fox, J. R. (1974) *Arming America: How the US Buys Weapons*, Cambridge, Mass: Harvard University Press.

Garrison, J. and Shivpuri, P. (1983) *The Russian Threat. Its Myth and Realities*, London: Gateway Books.

Halliday, F. (1983) *The Making of the Second Cold War*, London: Verso.

Halloran, R. (1982) 'Pentagon draws up first strategy for fighting a long nuclear war', *New York Times*, 29 May.

Holloway, D. (1984) 2 edn, *The Soviet Union and the Arms Race*, New Haven: Yale University Press.

Howard, M. (1973) 'The Relevance of Traditional Strategy', republished in M. Howard (1983) *The Causes of War and Other Essays*, London: Unwin Paperbacks.

Howard, M. (1976) 'The Strategic Approach to International Relations', republished in M. Howard (1983) *The Causes of Wars and Other Essays*, London: Unwin Paperbacks.

Howard, M. (1981) 'The Causes of Wars', republished in M. Howard (1983) *The Causes of Wars and Other Essays*, London: Unwin Paperbacks.

Johnstone, D. (1984) *The Politics of Euromissiles. Europe's Role in America's World*, London: Verso.

Joint Statement issued by President Carter and Prime Minister Margaret Thatcher, 18 December 1979.

Kaldor, M. (1982) *The Baroque Arsenal*, London: Andre Deutsch.

Khrushchev, N. (1971) *Khrushchev Remembers*, London: Andre Deutsch.

Kissinger, H. (1982) *Years of Upheaval*, London: Michael Joseph.

Komer, R. W. (1984) Address given to the Centre for Strategic and International Studies, Conference on 'The Future of NATO and Global Security', Brussels, January 1984, cited in D. Johnstone (1984) *The Politics of Euromissiles. Europe's Role in America's World*, London: Verso.

Lens, S. (1971) *The Military-Industrial Complex*, London: Stanmore Press.

McNamara, R. S. (1962) Address given at the University of Michigan, Ann Arbor, 16 June 1962, cited in Groom, A. J. R. (1977) 'The British Deterrent', in J. Baylis (ed) (1977) *British Defence Policy in a Changing World*, London: Croom Helm.

Prins, G. (ed.) (1982) *Defended to Death*, Harmondsworth: Penguin.
Rostow, E. (1984) cited in *Newsweek*, 16 July.
Sanguinetti, A. (1981) 'Les Interventions Militaires Françaises', *Tricontinental 1, La France Contre L'Afrique*, Paris, 1981, cited in D. Johnstone (1984) *The Politics of Euromissiles. Europe's Role in America's World*, London: Verso, p. 89.
Scheer, R. (1982) *With Enough Shovels. Reagan, Bush and Nuclear War*, London: Secker and Warburg.
Sims, D. (1970) 'Spoon-Feeding the Military – How New Weapons Come to Be', in L. Rodberg and D. Shearer (eds) (1970) *The Pentagon Watchers*, Garden City, NY: Doubleday.
Sunseri, A.R. (1977) 'The Military-Industrial Complex in Iowa', in B. F. Cooling (ed.) (1977) *War, Business and American Society. Historical Perspectives on the Military-Industrial Complex*, Port Washington, NY: Kennikat Press.
Tucker, R. W. (1980) 'American Power and the Persian Gulf', *Commentary*, November.
Weinberger, C. W. (1982a) cited in *Newsweek*, 15 November.
Weinberger, C. W. (1982b) *Annual Report to the Congress, Fiscal Year 1983*, Washington DC: US Government Printing Office.
Zuckerman, S. (1982) *Nuclear Illusion and Reality*, London: Collins.

6
The Effects of Nuclear War

The Evidence Available

Our knowledge of the likely effects of a nuclear war derives from a variety of sources including the known effects of the two nuclear bombs dropped on Hiroshima and Nagasaki in August 1945; information obtained from the various atmospheric nuclear tests carried out by the nuclear weapons states; data on the effect of accidental irradiation arising from, for example, the operation of nuclear power stations; and data on the effects of relatively low levels of radiation on human beings and the environment.

Hiroshima and Nagasaki

The nuclear weapon exploded on 6 August 1945 at a height of about 500 m above Hiroshima was a free-fall bomb using uranium-235 to produce a nuclear explosion with a force or yield of approximately 12,500 tons of TNT (12.5 kilotons). Almost everyone within 500 m of the point beneath the explosion (ground zero) was killed; about six out of every ten people within 2 km of ground zero died – most of them in the first 24 hours. It is not known how many people were in Hiroshima when the bomb exploded but the best estimate is that there were about 350,000. By the beginning of November 1945, 130,000 people (that is, about 40 per cent) were dead. This figure is almost certainly an understatement as the National Census taken in 1950 could not account for many thousands of people. Of those killed immediately, most were crushed or burned to death as an area of 13 sq km was reduced to rubble. All buildings within 2 km of ground zero were damaged beyond repair. Of the 76,000 buildings in Hiroshima, two-thirds were destroyed by fire. Commenting on the bombing on 8 August 1945, *The Times* stated: 'Destroyed is the word used officially, but it appears that "obliterated" might be a better word.'

The bomb dropped on Nagasaki on 9 August 1945 had a core of plutonium-239 surrounded by chemical explosives. Like the Hiroshima bomb it was exploded at a height of 500 m. The extent of damage from ground zero caused by this weapon, which is thought to have had an explosive yield of about 22 kilotons, varied considerably because the city is built on mountainous ground so that some areas were more shielded from thermal radiation and blast than others. Of the 270,000 people believed to have been in the city, 60,000 to 70,000 had died by the beginning of November; 7 sq km of the city were devastated and a quarter of the city's 51,000 buildings were totally destroyed and many more seriously damaged. William A. Lawrence, the science writer for the *New York Times*, commented of Nagasaki '12 hours after the bombing this city was still a mass of flames, it was like looking over the rim of a volcano in the process of eruption' (cited in CND, n.d.).

The bombs used on Japan were very inefficient weapons by today's standards, with small yields. In 1962 the Soviet Union exploded an H-Bomb with a yield of about 58 million tons of TNT (58 megatons), roughly equal to the power of 3,000 Nagasaki bombs. Although even higher yields can be obtained, bombs of this size are not really necessary. The largest city could be demolished by a bomb of about 10 megatons.

Assuming that the topographical features of the targets are similar, a bomb with a higher yield will destroy a larger area than a bomb with a smaller yield, but the relationship between the yield and area destroyed by blast is not proportional. A one megaton bomb will destroy an area only *four* times greater than a 125 kiloton bomb, even though the yield of the former is *eight* times that of the latter. Where large areas are to be devastated, a similar effect can be obtained by exploding a number of smaller warheads so that the areas destroyed by their blast overlaps. In fact, the prime objective of atomic weapons are now not cities but military targets, many of which are hardened to withstand blast pressure. To counteract hardening, and to allow for the inaccuracy of missiles, warheads with larger yields were introduced in the 1950s. As missiles and warheads have become more accurate, so it has been possible to reduce the size (and yield) of the warheads, thus allowing more independently targeted warheads to be carried per missile. The trend is thus towards smaller yields.

Nuclear Weapons Tests

Nuclear weapons have been tested since 1945. In October 1963 the USA, USSR and Britain agreed to ban nuclear weapon test explosions in the atmosphere or under water (Partial Test Ban Treaty). France and China refused to be bound by this treaty although France has adhered to it since 1975. Information on the accidental irradiation of people as a result of the various tests is not fully documented. During the 1954 'Bravo' test at Bikini Atoll, the wind veered and the crew of a small Japanese trawler, the *Fukuryu Maru* (*Lucky Dragon*) was affected by fall-out. All the crew suffered from radiation sickness and one of them died. The same test also deposited fall-out over three of the Marshall Islands, whose inhabitants also suffered from the effects of radiation fall-out. As late as 1983 an interim report to the US Department of the Interior from the Bikini Rehabilitation Committee noted that while the fish in the lagoon were safe to eat, home-grown food was still contaminated, and was likely to remain so well into the twenty-first century (*The Guardian*, 29 November 1983).

There has been an increasing amount of information coming out about the effects of the atmospheric tests carried out in Australia, the USA, and the Pacific on servicemen and other personnel directly involved in the tests, and on the inhabitants of areas directly affected by fall-out. A report in *The Sunday Times* (21 November 1982) pointed to the remarkably high incidence of cancers, birth defects, miscarriages and deaths, particularly from cancers, among people who lived downwind of the Nevada tests in the USA. A study by Dr Carl Johnson published in the *American Medical Journal* indicated that the Mormon inhabitants of St George and nearby communities in Utah have a cancer rate from almost double to 11.42 times that of Mormons in the rest of the state (*The Guardian*, 6 March 1983). There were 87 open-air tests of atomic bombs in the area between 1951 and 1962. An estimated 400,000 soldiers were used as guinea-pigs near the sites during the American atmospheric tests in the Pacific and in Nevada. *The Sunday Times* reported on 16 January 1983 a survey of 1,000 veterans of the American nuclear tests which showed that the incidence of cancer was twice that normal for men in the age-group, and that the incidence of sterility was twelve times higher than normal.

There is also evidence of radiation-related illnesses among servicemen who took part in the British nuclear tests in Australia and the Pacific. Initial evidence, reported in *The Guardian* on 17 March 1983, suggested that the number of leukaemias among the 13,000 or so servicemen who took part in the tests already appears, on incomplete data, to exceed the level which would normally be expected to occur. Although the Ministry of Defence dismissed the possibility of accidental over-exposure of servicemen through weapons being exploded unexpectedly close to the ground or turning out to have a higher yield than planned, there is a great deal of anecdotal evidence of accidental exposure (Tame and Rowbotham, 1982; *The Times*, 20 June 1984). According to reports in *The Times* (20 June 1984) some servicemen may have been as near as 1 mile from ground zero, and there is evidence that British servicemen were stationed in the open 4 km from ground zero. There is also evidence that some Aborigines were contaminated by fall-out and that some died (*Newsweek*, 21 May 1984). A Royal Commission is investigating these allegations.

It has also been suggested that fall-out from atom bomb tests in the 1970s may be responsible for the dramatic increase in young men suffering cancer of the testicles. Dr Karl Sikosa, director of the Medical Research Council's Ludwig Institute for Cancer Research at Cambridge points out that 'The pattern of incidence of testicular cancer matches the pattern of radioactivity from fall-out in the atmosphere with a lag of 20 years' (Sikosa, 1984).

Accidental and Low-Level Radiation

There is considerable evidence that accidents in nuclear plants and the effects of waste disposal have led to higher-than-average incidence of cancers. In Britain, an unexplained cluster of leukaemias in the Leiston area of Suffolk (the existence of which has been disputed) has been attributed to the proximity of the Sizewell A Nuclear power station in East Suffolk; the incidence of childhood cancer is said to be abnormally high near the Sellafield (Windscale) nuclear complex in Cumbria, the incidence of leukaemia at Seascale, a mile from the plant, being said to be ten times the national average; and the incidence of leukaemia in children living near the Winfrith nuclear reactor in Dorset is also said to be ten times the national average. Similar abnormally high incidences of cancer and cancer-related diseases have been observed in connection with nuclear generating and reprocessing plants in other countries. Levels of contamination in the Irish Sea are the highest of any sea in the world and have been linked with the plant at Sellafield. Studies of the incidence of leukaemia and lymphoma (cancer of the lymph glands) among adults and children in the West Coast of Scotland indicate an increase in the frequency of these cancers since 1958 when the plant at Sellafield began operating (*The Sunday Times*, 13 November 1983).

The claims of the nuclear power industry to be a safe industry have been strongly challenged. The belief, once widely held and still encouraged by the nuclear industry, that there is a threshold beneath which exposure to radiation is acceptable, has been vehemently challenged. Caldicott (1980, p. 19) states: 'Whether natural or human-made, all radiation is dangerous. There is no "safe" amount of radioactive material or dose of radiation.' Bertell (1985, p. 45) agrees, but points out that a 'permissable level, based on a series of value judgements, must be set', thus weighing up the risks and balancing them against the benefits to be derived. She draws attention to the problems of making such judgements and the failure of governments to try to measure the effects of radiation on the population: 'Continuance of present government neglect and unconcern is at best irrational and at worst genocidal' (Bertell, 1985, p. 62).

The Physical and Medical Effects of Nuclear Weapons

The physical effects of a given weapon vary according to the height above the ground at which it is exploded. If the height is such that the fireball touches the ground, the explosion is said to be a 'ground-burst'. For any height above this it is an 'air-burst'.

Thermal Radiation and Initial Nuclear Radiation

When an atomic bomb is detonated, a fission chain reaction is triggered and sustained in order to produce a large amount of energy in a very short time. The effect of the heat generated by the reaction on the surrounding air causes a very powerful explosion. The size of the explosion depends on the amount of fissionable material contained in the bomb, the material used, the design of the bomb, and its efficiency.

When the weapon is exploded above the ground, the first noticeable effect is a blinding flash of intense white light which is emitted from the surface of the 'fireball' – a roughly spherical mass of very hot air and weapon residues which develops around the exploding weapon and then expands. The radius of the expansion depends on the size or yield of the bomb. A bomb of the type dropped on Hiroshima or Nagasaki produces a fireball of about 200-m radius in 1 second. As the bomb explodes, intense ionizing radiation consisting primarily of neutrons and gamma rays will be emitted. Up to 1½ miles away an unprotected person will receive a lethal dose, although anyone as close as this would be killed by blast and heat. Generally, except for bombs with small yields, the effects of initial nuclear radiation can therefore be ignored. During this period and for some time after the fireball also emits thermal radiation in the form of light and heat. As the fireball cools the thermal radiation dies away. At the same time, the fireball rises through the air as a bubble through water until it reaches the top of the troposphere where it spreads out into the familiar mushroom shaped cloud. By then about one-third of the explosive energy will have been released as heat. In the area of the fireball everything will evaporate or melt. At some distance from the explosion the heat will cause burns on exposed skin and ignite fires. Third-degree burns where parts of the skin are destroyed occur at 2 km from the explosion for Hiroshima-sized bombs; second-degree burns causing pain and blisters will occur at 3 km. Most people within 2 km of the explosion who are directly exposed to it (i.e. who suffer third-degree burns) will be killed by the thermal radiation. The ignition of other materials will cause fires that, in certain conditions, may cause fire storms, which add to the casualties.

The number of burns caused by direct exposure to the heat of the fireball depends on the yield of the bomb, the height of detonation and the weather conditions. In clear visibility (19 km) with no cloud or snow cover, the distance from ground zero at which people would be killed or injured, measured in km, is shown

Table 6.1 Distance from ground zero (km) at which people would be killed or injured by burns after a nuclear attack, assuming clear visibility

Explosive yield (Megatons)	Ground-burst		Air-burst	
	Injured	Killed	Injured	Killed
0.15	6.1	4.6	8.4	6.4
0.5	9.6	7.5	13.2	10.3
1.0	12.2	9.7	19.1	14.7
10.0	23.6	19.8	47.7	38.2

Source: Openshaw et al. (1983) p. 118.

in table 6.1. Here death is said to be caused by a radiant exposure of 6.7 calories per sq cm, a level sufficient to set light to newspapers or synthetic fabrics.

Blast

The air blast created by a nuclear explosion carries about half the explosive energy and travels much more slowly than the various forms of radiation. Depending on the size of the bomb, the blast wave reaches the 1 km circle in about 1 or 1½ seconds, and in 5 or 6 seconds has expanded to 3 km. The blast wave comes as a shattering blow followed by hurricane winds travelling out from the centre of the explosion. At 1½ km, wind speeds for a 12.5 megaton bomb will be about 336 km per hour (about three times the force of a 'full gale'). Trees will be uprooted, telephone poles blown down and virtually all buildings will be demolished. Anyone in the open will be swept off their feet. People will be injured, crushed or killed by falling or blown objects. Within a 1½ km radius the blast wave from a 12.5 megaton bomb will kill virtually everyone outside or inside ordinary buildings. The primary blast damage will occur during a few seconds. The effect of the blast wave may be modified by topological features. Table 6.2 provides information on the distance from ground zero (in km) at which a given percentage of people are killed or injured as a result of blast. The effect of the blast wave travelling out from the explosion is to

Table 6.2 Distance from ground zero (km) at which maximum over-pressures would be experienced from blast of 1, 2, 5, and 12 pounds per square inch (psi), and percentages of population killed or seriously injured by blast, in different over-pressure ranges

Maximum over-pressure (psi)	Ground-burst				Air-burst			
	12	5	2	1	12	5	2	1
Explosive yield (Megatons)								
	km	km	km	km	km	km	km	km
0.15	1.5	2.4	4.1	6.2	2.0	3.5	6.3	9.4
0.5	2.3	3.6	6.2	9.3	3.0	5.2	9.5	14.0
1.0	2.9	4.5	7.8	11.8	3.8	6.6	12.0	17.7
10.0	6.3	9.8	16.8	25.4	8.2	14.2	25.8	38.1
Over-pressure range (psi)	<12	5–12	2–5	1–2	<12	5–12	2–5	1–2
Killed (%)	98	50	5	0	98	50	5	0
Seriously injured (%)	2	40	45	25	2	40	45	25

Source: Openshaw et al. (1983), pp. 122, 123. Based on Glasstone and Dolan, (1977), p. 115, and Office of Technology Assessment (1980), p. 19.

create behind itself an area of low pressure. Eventually the air, in the form of a violent wind, will flow back to fill this.

The range of blast effects are greater for air-burst weapons, so they would be used if the objective is to cause the maximum amount of damage to unprotected targets such as cities or large industrial areas. Ground-burst bombs create a much higher level of blast pressure at ground level. Some of this energy is transmitted into the ground to produce an 'earthquake' effect which damages underground structures and forms a crater.

Fall-out

In a ground-burst or near surface burst, vast quantities of dirt and debris are gouged out of the crater and sucked up into the fireball.

The debris is irradiated and it is these radioactive materials that are later swept up into the sky by the inward rushing wind and then subsequently deposited as fall-out. Since ground-bursts are the way to destroy hardened targets, they are likely to occur as a result of the US and Soviet counter-force strategies.

Some of the fall-out returns to earth within hours or days. This is called 'early fall-out'. Smaller particles may be carried up into the stratosphere where they can remain for months or even years. This 'delayed fall-out' may be dispersed on a global scale. With ground-bursts about 40 per cent of the fall-out remains air-borne for long periods. Fall-out is the most important longer-term effect of ground- and near-surface bursts. The early fall-out is deposited on the areas immediately around and down-wind from the explosion, the area affected possibly stretching down-wind for hundreds of kilometres.

The amount of local fall-out depends on the yield of the bomb, the fraction of the yield obtained from fission reactions, and the nature and materials of the ground surface. The direction and speed of the wind, as well as the presence of rain, affects fall-out, so that the pattern of radiation contamination on the ground can be very irregular with local areas of especially high contamination ('hot spots'). Studies of the effects of nuclear weapons usually show fall-out drifting down-wind of the point of explosion in an idealized plume although the actual drift will be affected by terrain, wind direction, and so on, and will be much more irregular.

With the lapse of time, the level of radioactivity of the fall-out decreases. The rate of decay is sometimes referred to as the 'seven-tenths' rule – that is, the intensity of the activity reduces by a factor of 10 as the time lengthens by a factor of 7. A dose rate of 100 rads per hour will fall to 10 rads after seven hours and to 1 rad after 49 hours. At ground level, however, the radiation dose rate will rise in the days after the explosion as fall-out accumulates, before dropping as the process of radioactive decay begins to lower the accumulated dose rate faster than any increase caused by additional fall-out.

What constitutes a lethal dose of radiation is subject to dispute. The absorbed dose is now expressed in units called the gray (symbol Gy). This unit has replaced the rad (1 Gy is equal to 100 rads). It is clear that comparatively low whole-body doses of radiation commonly cause anorexia, nausea and vomiting. Above a dose of 1 Gy bone marrow damage occurs. Without treatment a dose of 4–4.5 Gy is thought to result in the death of half the exposed young, fit, adult population, while survival is unlikely if the dose is above 6 Gy. The lethal dose at which half of the population dies (called the LD 50) varies depending on the length of time over which the dose of radiation is received, whether the whole body or only part of the body is exposed, and the age of the victim. Children and old people are particularly susceptible. In the aftermath of a nuclear war, the effects of cold, hunger and disease may lower the LD 50 to 3.5 Gy.

Three main varieties of radiation sickness are recognized. The bone-marrow form occurs after exposure of the whole body to 1.5–4 Gy. There is a short-lived period of lethargy and nausea over 1 to 2 days, then a symptom-free period of about 10 days, followed by a period of about 2 weeks in which the victim suffers fever, skin haemorrhages and mouth ulcers. Hair loss may occur where the dose is in the range 3–4.5 Gy. The depression of white blood cells makes the victim susceptible to other infections and to spontaneous haemorrhages. Some individuals may die; most will probably recover.

The gastro-intestinal form of radiation sickness occurs in the range of 4.5–10 Gy. The victim feels generally unwell on the first day. From about the seventh day there is massive, bloody diarrhoea with loss of body fluids, fever and the risk of septicaemia. If the victim is still alive at the end of 3 weeks, then he or she will continue to suffer from the bone-marrow form of radiation sickness. Above 6 Gy almost all people would die. The third form of radiation sickness, the central nervous system form, is encountered with doses of 10 Gy or more. In its mildest form the victim suffers lethargy, unsteadiness, convulsions, coma and death within days. In its more severe form, coma, convulsions and death occur within hours.

Other effects which are noticeable include burns on the skin caused by Beta radiation from fall-out falling on the skin; the retardation of fetuses including, for example, damage to brain growth in children whose mothers are less than 15 weeks pregnant and who receive a dose of 2 Gy. Fertility is likely to be impaired. A dose of 2.5 Gy will produce sterility in males for 3 years or longer. Doses of 1–2 Gy produce temporary sterility in women. Doses of about 4 Gy will cause permanent sterility in both men and women.

The longer term effects of irradiation arise as late consequences of exposure to initial nuclear radiation or early fall-out, or from continued exposure to long-lived radioactive particles which may enter the body through food or drink, by inhalation, or (rarely) through the skin. The incidence of leukaemias, and solid cancers of the thyroid, bone, female breast and lung, is increased. The incidence of leukaemia peaks about 6 to 7 years after exposure; other forms of cancer have a latent period of about 25 years. The increased risk of dying from a radiation-induced malignant tumour is unknown, but there is no doubt that the incidence of cancers in populations that have been exposed to radiation is higher than that for non-exposed populations. Early estimates based on atomic bomb survivor data and human X-ray research suggested that the dose at which the incidence of leukaemia and solid cancers doubled were 1 Gy and 3–4 Gy respectively. However, work by Dr Thomas Mancuso of the Public Health Department of the University of Pittsburg on the health records of workers in the USA has suggested that the 'doubling dose' stands at 3.6 rads per lifetime for bone marrow cancers and 33 to 38 rads per lifetime for other forms of cancer. In other words, a civilian nuclear power worker exposed to a radiation dose of 5 rads per year would be exposed to more than a doubling dose for leukaemia each year and to a doubling dose for cancer every 7 to 7½ years (cited in Caldicott, 1980, p. 22). As indicated earlier, no dose of radiation is so low that there is no increased risk of cancer being induced.

There is also a risk of genetic abnormalities. Such effects would not become apparent for many generations. An increased number of chromosonal abnormalities in cultures of white blood cells have been noted in survivors of the Japanese atomic bombs 20 years later. As Dr Helen Caldicott (1980, pp. 16–17) points out, almost all geneticists now agree that there is no dose of radiation so low that it produces no mutations at all.

As indicated earlier, the effects of radiation on human beings vary considerably between individuals. For this reason the effects are usually expressed in terms of the percentage of those exposed to a given dose who are likely to die or fall ill as a result. Most sources give the LD 50 at about 450 rads (4.5 Gy). This is the figure used by Glasstone and Dolan (1977), and in the study by the US Congress's Office of Technology Assessment (1980). The confidential *Guidance Notes for Doctors Teaching Mass Casualty Care* issued by the British Department of Health and Social Security suggests that anyone who has probably had a dose of more than 350 rads (3.5 Gy) should have 'a low priority ... and be treated expectantly' (DHSS, 1979, cited in Campbell, 1982, p. 387). This is the euphemism used for those who are expected to die. Generally speaking, the evidence on which the figures are based is rather scanty. The actual doses received by the many victims of radiation in Hiroshima and Nagasaki were not measured, while data from other sources (e.g. accidental irradiation, intensive hospital treatment, experiments on animals) is insufficient or not wholly applicable to calculations of human casualties in a nuclear war.

Psychological Effects of Nuclear Explosions

In addition to physical injuries and radiation sickness, survivors would be subject to immense psychological stress. These effects are difficult to predict with certainty but were noticeable in the aftermath of the Hiroshima and Nagasaki bombings. Many people would be incapable of organized activity. In addition to the shock engendered by bereavement, physical suffering and a devastated landscape, there would also be the psychological problems of dealing with fall-out which, since it cannot readily be perceived by the physical senses, would be likely to cause considerable anxiety. The disruption to communication systems would mean that survivors would not know what had happened to their relatives and friends, and this would add to their stress. Despair, apathy, depression, memory lapses and confusion have been noticed in the survivors of natural disasters. These psychological effects can last for weeks if not months or years.

British Home Office Assumptions affecting Casualty Figures

There is a marked discrepancy between casualty figures produced by independent scientists in Britain using data derived from US sources, notably the standard work by Glasstone and Dolan (1977) on the effects of nuclear explosions, and those prepared by Home Office scientists. These discrepancies, which have resulted in an energetic debate, arise in three areas:

1 The Home Office takes no account of deaths or injuries caused from thermal radiation on the grounds that people will take cover when they hear the attack warning. This is not too serious since deaths from burns are a relatively small proportion of total deaths (Openshaw et al. 1983, pp. 198–9).
2 The Home Office seriously underestimates the peak overpressures caused by blast at given distances from ground zero. This is particularly true of airbursts (Openshaw et al. 1983, pp. 199–205).
3 The Home Office has established an Operational Evaluation Dose (OED) which provides a standard for the dose which civil defence workers and others could be expected to take when working out of shelters under conditions of heavy radiation and survive. The OED is defined as the actual dose received less 150 rads less 10 rads per day since exposure started (subsequently revised to 200 rads and 15 per day). These deductions are made on the grounds that it is only above the base rate of 150/200 rads that the risk of death from exposure rises significantly, and that the rate of 10/15 rads per day is the rate at which the body is said to recover from radiation effects as cells repair themselves. However, even at 150 rads 1 in 12 healthy young adults die, and at this level of exposure severe bone marrow damage results. There is also no data showing that the body can repair itself indefinitely at 10 rads or more per day. Finally, the longer terms risks of cancer and leukaemia increase steadily with the accumulated dose of radiation, starting from the first rad (Openshaw et al. 1983, pp. 208–9; Professor Patricia Lindop, 1981, cited in the British Medical Association, 1983, p. 88).

The Effects of Nuclear War on Britain

A number of studies have been done on the likely effects of 'limited' nuclear attacks on British cities. These include a study by Qasrawi et al. (1982) on the effects of an attack on the West Midlands and a study by Greene et al. (1982) of the effects of an attack on London. There are, however, very few people who believe that a nuclear attack on Britain would be limited to a few bombs. On the other hand there is no way of knowing exactly how many weapons are targeted on Britain. The introduction to the

Home Office's *Training Manual for Scientific Advisers* (Home Office, 1977) suggested an attack of about 200 megatons and this was also the level used in the Operation 'Square Leg' training exercise. Evidence given by a Home Office spokesman to the Working Party of the British Medical Association's Board of Science and Education, indicated that the pattern of attack is 'impossible to predict' (BMA, 1983, p. 23). However, Britain's independent nuclear deterrent, together with the fact that the country is a staging post for US forces, makes it an important target. The Home Office spokesman said that the 'Square Leg' scenario of a 200 megaton attack on about 80 targets 'in no way reflects our assessment of the possible scale of attack the Soviet Union might launch against this country if a war in Europe developed into a nuclear war' (BMA, 1983, p. 23). The *Training Manual for Scientific Advisers* says that 'It... seems likely that any initial nuclear strike would be a massive one aimed at making the country attacked totally ineffective militarily, politically and industrially. In particular, the means of nuclear retaliation would be primary target' (Clayton, 1978). Openshaw et al. (1983, Appendix 1) provide a detailed list of potential targets in Britain, by grid square reference (1983, pp. 242–69). Another list is provided by Duncan Campbell (1982, pp. 424–6). As he points out, 'The weapons exist, and the targets exist' (p. 438). Examining 'the extraordinary biases which have featured in the casualty and damage assessments made by the Home Office', Campbell argues that the Home Office deliberately underestimates the likely level of a nuclear attack on Britain (p. 439).

The deployment of American ground-launched cruise missiles at Greenham Common near Newbury and at Molesworth near Huntingdon could also influence the level of an attack. In 1981 Geoffrey Pattie, Secretary of State for Defence (Air Force), admitted in a written Parliamentary Reply that 'more than 1,000 megatons would be needed to destroy the ground-launched cruise missiles once they are dispersed' in an area up to 100 miles from their bases (BMA, 1983, p. 24). On the evidence available to it the British Medical Association's Working Party concluded that an attack of 150–200 megatons seemed likely, but that this would be increased by three or four times if cruise missiles were deployed (BMA, 1983, p. 25). Higher levels of attack might occur, possibly up to 3,000 megatons.

Although no one knows exactly what might happen in a nuclear war, it is possible to estimate the likely effects of a war given particular assumptions about the size and nature of an attack. The methods of calculation cannot be precise. They depend on extrapolations from the effects of the bombs dropped on Hiroshima and Nagasaki, which were small by today's standards, and on theoretical models developed on the basis of these bombs and the atmospheric tests carried out in the 1950s and 1960s. No one knows exactly what the effect of a large number of explosions will be because large weapons have never been exploded over cities or in large numbers simultaneously. Assessment of the likely effects also has to take into account the amount of warning given to the population to prepare for an attack, the weather conditions, whether the attack occurs at night-time or in the day, population density, topology and so on. However, the effects that cannot be calculated are likely to be as important as those which can be. For example, radiation sickness lowers the body's resistance to other infections. Under these conditions injuries and burns are more likely to prove fatal.

Openshaw et al. (1983, pp. 73–92) analyse 12 attack scenarios (see table 6.3). In evaluating their scenarios it is worth noting that they 'do not believe that any attack smaller than Attack H is plausible'. In addition some important targets will almost certainly be double- or triple-targeted. The 5 million people who remain uninjured after Attack K 'are mainly in the Scottish Highlands, the Welsh mountains and other such lightly populated regions with few targets' (Openshaw et al., 1983, p. 142). The authors point out that heavy industrial targets are only included from Attack H onwards: 'It is simply not the case... that a strictly "counterforce" attack, with relatively low-yield weapons, would result in only a small number of casualties' (p. 142). The only recent information on casualty levels available from the Home Office is in the form of a paper by S.F.J. Butler. He states that an attack with 179 weapons with a total explosive yield of 193 megatons directed at military targets and cities is estimated to give total deaths in the range of 17–26.9 million (Butler, 1981, summarised in BMA, 1983, p. 53). The average yield of the bombs is just over 1 megaton.

Openshaw and Steadman (1982) studied the pattern of attack envisaged in the Home Office 'Square Leg' Exercise and concluded, using American Office of Technology Assessment data on the effects of bombs, that total initial deaths would amount to 26.7 million, with another 6.8 million people seriously injured (see BMA, 1983, p. 54 for a summary). The 'Square Leg' exercise assumed 125 weapons with a total megatonnage of about 200, and an average yield per bomb of over 1.5 megatons each. They then went on to assume an attack of 200 1-megaton weapons designed to kill as many people as possible. Their computer-based model showed that 33.1–43.3 million people would be killed, but this took into account only those killed by blast, and not those killed by thermal radiation (burns) or fall-out.

Most Home Office estimates have assumed that, in general, bombs with a large yield (of 1 megaton or above) will be used. This restricts the number of targets but means that the area devastated around each target will be larger (say 10–12 miles for a 1-megaton bomb) than would be the case if smaller weapons were used. Some independent British researchers (Greene and Turok) suggested in a study which was used by the Campaign for Nuclear Disarmament to counter the Home Office 1982 Hard Rock exercise, that, with the development of more accurate missiles such as the SS-20 with its multiple independently targetable warheads, the number of warheads could rise although the total megatonnage dropped on Britain would remain about the same. This would increase the area of devastation. They assessed the likely effects of an attack primarily on military targets with some industrial targets, but using bombs with a smaller average yield. They envisaged an attack with 340 weapons with a total yield of 222 megatons (i.e. an average yield of 0.65 megatons per bomb). This gave a much higher death rate, of 38.6 million, with some 4.3 million seriously injured. This study was developed and used by Openshaw et al. (1983).

Everyone is agreed that the level of deaths in the immediate aftermath of a nuclear war could be reduced if adequate shelters were provided for the population. The protection afforded by buildings and shelters is called the Protective Factor (PF). A PF of 3 means that the dose received by someone inside a particular shelter would be one-third of the dose received by someone standing outside that shelter. The Home Office estimates that a shelter in the 'fall-out' room of an undamaged bungalow has a PF of from 5 to 10; one in a semi-detached two-storey house with 11-inch cavity brick walls would have a PF of 40 and so on. The improvized garden shelters described by the Home Office (1981)

Table 6.3 Attack scenarios in Openshaw et al. 1983)

Scenario	Number of warheads	Total yield (megatons)	Persons killed (millions)	Persons seriously injured (millions)	Killed and seriously injured as % of total population
A Nuclear forces and associated C^3I centres	75	42.5	6.4	2.7	17
B Scenario A plus military/civilian airfields, air defence and ASW bases, important signals intelligence and military radar stations, second-level nuclear targets	160	87.0	13.3	6.0	36
C A–B plus all other major nuclear and conventional military bases and some civilian ports and military related nuclear plants	242	116.9	20.1	6.2	49
D A–C plus small airfields, military bases and communication centres	286	129.45	20.8	6.4	50

E	A–C plus oil refineries and terminals, North Sea gas terminals, other ports, nuclear power stations, Bank of England	293	163.75	25.6	6.4	59
F	A–E plus large conventional power stations, some important transport links, etc.	315	185.75	28.4	7.1	66
G	A–F plus major ordnance factories, some major military industries	348	213.65	35.4	5.4	76
H	A–G plus major chemical and other heavy industrial plants	342	219.2	37.5	5.1	79
I	A–H plus increasing	392	257.65	39.9	4.4	82
J	number of major plants	441	306.65	42.9	3.6	86
K	and, in K, some urban-industrial targets	485	347.65	45.3	3.0	90

Source: Adapted from Openshaw et al. 1983, pp. 74–77, 140.

are said to have a PF of over 40; a shelter based on the standard World War II 'Morrison' shelter with a steel table surrounded by bricks, sand or other heavy materials would have a PF of 70. Pre-fabricated steel shelters sunk into the ground and covered with earth, similar to the World War II 'Anderson' shelter, are said to have a PF of not less than 200. Home Office models assume an average PF of over 20.

In its studies the Home Office generally assumes that the entire population are able and willing to shelter by blocking doors and windows and building 'inner refuges', and that they subsequently stay in these for at least two weeks and possibly longer. This may not be the case. More importantly, the Home Office makes no allowance for the effects of blast damage on buildings which would reduce the protection against radiation. Generally speaking, allowing for the amount of time people spend inside and out of doors under normal conditions, the PF afforded by an unmodified building is about 3. Openshaw et al. (1983, p. 133) suggest that for buildings suffering less than 1 psi of blast over-pressure, a PF of 8 seems reasonable. In areas suffering higher levels of blast damage, this will be diminished. A PF value of 1 (i.e. the same value as for being outdoors) is assumed for areas suffering a blast over-pressure of more than 5 psi. Any damage to the windows, roof or fabric of the house, creating openings through which fall-out could blow, will reduce the PF factor. The papers by Openshaw and Steadman (1982 and 1983), and the studies by Greene et al. (1982) and by Openshaw et al. (1983) show that damage to housing will reduce the PF factor and have an appreciable effect on the number of casualties. In addition, of course, any attempt to leave the house or shelter for any purpose (to look for food, water, relatives or whatever) will increase the dose received by the individual, and hence affect the level of deaths from radiation sickness.

In a critical study of the Home Office's Square Leg exercise, Openshaw and Steadman estimated that deaths from blast would amount to 23.6 million with a further 17 million casualties from fall-out if the population was totally unprotected. They went on to estimate that if the population was protected with an average PF of 4, then 10 million of their worst-case 40.6 million mortalities would live; a PF of 10 would reduce deaths by 13.8 million; and a PF of 40 would reduce deaths by 16.6 million (BMA, 1983, p. 79). There is a general agreement that the number of people saved from death as a result of fall-out by investing in shelters with a PF of more than 10 is relatively small. The next step in the saving of lives would be to construct blast-proof shelters.

Nuclear War in the Northern Hemisphere

The Royal Swedish Academy of Sciences (1982) studied the possible result of a nuclear war between the Superpowers in Europe and Asia in which about 4,970 warheads with an explosive power of 1,941 megatons are used against cities with a population of more than 500,000; another 3,136 warheads (701 megatons) are used on key industrial sites, energy supplies, and mineral resources; and a further 6,620 warheads (2,960 megatons) are used against military targets. Most of these warheads are used against targets in the northern hemisphere. Of an urban population of nearly 1.3 billion in the hemisphere, the study estimates that about 750 million would be killed outright and some 340 million would be seriously wounded. Many of the 'initial' survivors would later die of the latent effects of radiation and infectious diseases compounded by food and water shortages, lack of health care and lack of shelter and heating. The Academy's report, first published in 1983 in its journal *Ambio*, shows that a nuclear war would have such disastrous consequences that no side could win. The industrialized societies of the northern hemisphere would be destroyed. The collapse of the world's economic system would lead to the collapse of society in the southern hemisphere, and to widespread death through famine and disease. The environmental system on which man relies would be massively damaged and society as we know it cease to exist.

In the immediate aftermath of a nuclear war the survivors will need an adequate supply of uncontaminated water. Those suffering from vomiting, diarrhoea or fever will need more water than usual. Any widespread nuclear attack would quickly disrupt the distribution system for domestic and industrial water and much of the sewerage system. Covered water will be protected from fall-out but the amount stored is likely to be limited. In regions directly exposed to local fall-out, rainwater will be highly radioactive and unfit to drink. Water drawn from reservoirs, ponds and rivers will be contaminated by radioactive isotopes which dissolve in water. Other radioactive material is likely to fall to the bottom so that running water may be safe to drink provided the bed remains undisturbed. Equally serious may be the pollution of water supplies following the seepage of sewage from damaged sewage and waste disposal systems and contamination from human and animal bodies.

So far as food is concerned, the complicated global food-distribution system is likely to be severely disrupted by a nuclear attack. The British Government assumes that people will be able to build up a personal 14-day food stock before an attack. This may not be possible even if there is sufficient warning of an attack, given likely shortages following panic buying. Contamination, storage and distribution all pose problems. The inevitable collapse of the electricity supply industry will mean that frozen stocks will go bad. Some stockpiles will also be destroyed in the attack. Imports from abroad cannot be expected for months or years after the attack.

Agriculture, at least in the interim, would be unlikely to meet all the needs of the survivors. Livestock would be susceptible to radiation sickness so it might be worth killing off surviving animals and salting them, assuming that sufficient supplies of salt can be made available. Where cereal and pulses are available, their harvesting, distribution and processing would pose difficulties. Plants are also sensitive to radiation, although different crops are more or less resistant and the sensitivity varies depending on the stage of growth reached by the plant. The concentration of radioactivity into the food chain could have important effects for animal and human survivors. This would continue to pose dangers several years after an attack. Insects are generally much less susceptible to radiation and pests could multiply. The destruction of industry would affect the production of pesticides, as well as chemical fertilizers. In developed countries farming is highly mechanized and heavily dependent on fossil fuels, electrical power, machinery, equipment and spare parts. The destruction of this infrastructure would mean that farming would have to revert to traditional methods. However, this would take time. Even if sufficient old-fashioned implements were available, the techniques of using them have been lost. With many people facing starvation the retention of seeds for the new crop would pose problems of control.

Power, transport and communication systems would be severely disrupted. Weakened by radiation sickness, illness and lack of food, the survivors' ability to perform physical work is likely to be impaired. Psychological problems will also affect peoples' ability and will to work. There will be severe shortages of skilled workers. The institutions upon which the modern economy is founded – financial institutions, banks, government agencies, companies – will have little meaning. Barter is likely to replace money as the medium of exchange. Summarizing the economic consequences of a nuclear war in the northern hemisphere for the study undertaken by the Royal Swedish Academy of Science, Yves Laulan (1982, p. 146) concluded:

> In a post-nuclear war society economies would be devoted to managing very scarce resources indeed. The survivors would find themselves confined to specialised societies with highly planned economies contained inside rather authoritarian regimes. War economics geared to sheer survival, with no regard for growth, would be the dominant configuration. In other words, we would be returning to a sort of economic dark ages: inward looking, and locked up in an autocracy that would be primarily agricultural.

One of the *Ambio* papers, prepared by two scientists, Paul J. Crutzen of the Max Planck Institute of Chemistry at Mainz and John W. Birks of the University of Colorado, suggested that as a result of 'the many fires ignited by the thousands of nuclear explosions in cities, forests, agricultural fields and oil and gas fields' (Crutzen and Birks, 1982, p. 74), vast quantities of smoke would rise into the atmosphere. The effect would be to screen out a large fraction of the solar radiation for many weeks, depriving plants and phytoplanckton of sunlight. Crutzen and Birks drew attention to the connection between the dust and debris injected into the atmosphere following the impact of a large meteor with the earth and the climatic change that led to the widespread and massive extinctions which took place at the Cretacious-Tertiary boundary about 65 million years ago. The effect of smoke and dust caused by nuclear explosions would, they suggest, cut sunlight and lead to a sharp drop in temperatures. Agricultural production would be almost totally eliminated in the Northern Hemisphere so that no food would be available for the survivors. Subsequently, as the sky began to clear and the sunlight began to come through again, the Earth would be subjected to increased levels of ultraviolet radiation. The reason for this is that the explosion of high-yield nuclear weapons would result in the formation of oxides of nitrogen which when injected into the stratosphere, would react with and reduce the quantity of ozone in the stratosphere. The ozone layer protects the earth's surface from much of the harmful ultraviolet radiation which comes from the sun. A 1975 report by the US National Academy of Sciences calculated that a large-scale nuclear war would reduce the quantity of ozone in the stratosphere by from 30 to 70 per cent. As a result, people would suffer sunburn from even short periods in the sun and could be temporarily blinded if they were not wearing protective clothing and glasses. They could develop skin cancer. Some plant crops might be affected, while single-cell organisms (phytoplankton) in the shallow waters of the oceans, upon which other organisms rely for food, might be depleted or killed outright. Animals might well be blinded. The overall effects, while impossible to predict with certainty, are likely to be serious, making survival more difficult.

Nuclear Winter

Further indications of the possible effects of a nuclear war come from two other sources: the effect of volcanic eruptions on the Earth's climate, and the effects of prolonged dust storms on the climate of Mars. Evidence obtained from Mariner 9, which reached Mars in 1971 during a prolonged dust-storm, showed that the effect of the storm was to raise the temperature of the Martian atmosphere and to cool the surface temperature. These observations initially led a group of scientists – commonly known as the TTAPS group, to try to find out what happens to the climate of the Earth when a large volcanic eruption throws particles up into the stratosphere (Turco et al., 1983). The effect of the eruption of El Chinchón in Mexico in 1982 and of Tambora in 1815 on climatic conditions was noticeable. The volcanic explosion at Tambora in Indonesia led to an average global temperature decline of 1° Centigrade and resulted in the 'lost summer' of 1816 in Europe. In an Ice Age, a typical long-term temperature decline from pre-existing conditions is about 10°C.

The TTAPS group then began to investigate the likely effects of a nuclear war in which large amounts of dust would be thrown up into the atmosphere. At about the same time Crutzen and Birks' *Ambio* (1982) study was published. Drawing on this, the TTAPS group adapted the computer models they had developed to look at the effects of dust to take account of smoke as well. They discovered that smoke will have a greater effect on the climate than dust. Black oily smoke from burning cities and from oil fields absorbs the Sun's rays, while very fine particles of dust scatter the light without absorbing it. Moreover, the particles of smoke will in general be smaller than the dust and so will, provided they are thrown up high enough, stay in the atmosphere longer than dust.

The next stage in the TTAPS study was to model the effect of various nuclear war scenarios. The group chose a 5,000-megaton exchange, that is about one-third of the world's total stockpile, as their 'baseline' or 'reference' scenario. Other cases ranged from a total yield of 100 megatons to a 'future wars' scenario involving 25,000 megatons. Generally, a modest fraction of the total yield (15–30 per cent) was assigned to urban targets, but in one case 1,000 × 100 kiloton bombs (100 megatons) were assigned to cities only.

The results of these investigations were reported at the conference 'The World after Nuclear War' held in Washington in October–November 1983, and subsequently published (Turco et al., 1983). They showed that while oceanic temperatures in the Northern Hemisphere would fall by only a few degrees, temperatures in the continental interiors would drop dramatically following a 5,000-megaton exchange to a minimum land temperature of −23°C after three weeks. Subfreezing temperatures would persist for several months. Even less severe exchanges (of 1,000–3,000 megatons) would cause temperature drops of from 5 to 10°C, and turn summer into winter. Moreover, a 100-megaton cities-only attack would be sufficient to produce a 2-month interval of subfreezing temperatures, with a minimum temperature of about −23°C

The biological effects of the TTAPS scenario were studied by a group of about 40 scientists at a meeting held at Cambridge, Massachusetts, in April 1983. Their findings were also reported to the Washington Conference and subsequently published (Paul

Ehrlich et al. 1983). The biologists found that the sudden fall in temperature coupled with low light levels (which affects photosynthesis), the effects of radiation, and the effects of toxic smog caused by fires, could destroy the biological support systems of civilization, at least in the Northern Hemisphere, and possibly severely disrupt or destroy the tropical forests and 'lead to the extinction of most of the species of plants, animals, and microorganisms on the Earth' (p. 1299). Survivors in the Northern Hemisphere 'would face extreme cold, water shortages [because water would be frozen and fuel supplies for melting ice very limited], lack of food and fuel, heavy burdens of radiation and pollutants, disease, and severe psychological stress – all in twilight and darkness' (p. 1299). If, as is possible, darkened skies and low temperatures spread over the whole globe, 'a severe extinction event could ensue, leaving a highly modified and biologically depauperate Earth' (p. 1299). Although *homo sapiens* would probably not be forced into extinction immediately, 'it is clear that the ecosystem effects *alone* resulting from a large-scale thermonuclear war could be enough to destroy the current civilization in at least the Northern Hemisphere' (p. 1299).

Subsequent models produced by other researchers have tried to redress some of the weaknesses of the TTAPS model (for example, those developed by Mike MacCraken of the Lawrence Livermore National Laboratory, USA, Dr Vladimir Alexandrov, Director of Climate Modelling at the Computer Centre of the Soviet Academy of Sciences, and Curt Covey of the US National Centre for Atmospheric Research at Boulder, Colorado). All have broadly confirmed the TTAPS findings. A report commissioned by the US Department of Defense and prepared by 18 specialists appointed by the National Research Council agreed that a nuclear war between the superpowers which occurred during the spring or summer in which about half their stockpiles of weapons were used could pollute the atmosphere with smoke and dust for 6 to 20 weeks and would lead to sharp falls in temperature in North America and Eurasia, the effects of which could become catastrophic within days and might last for months in the northern temperate zones, with almost total loss of light (National Research Council, 1985).

The future in a post-nuclear world is far from encouraging. No aspect of life would remain untouched. G.M. Woodwell, reviewing for the Royal Swedish Academy of Science *Ambio* study the effect of radiation on natural plant and animal kingdoms, and the enormous ecological damage of a nuclear war, concludes: 'There would be survivors, as there would be surviving plants, other animals, and here and there, an untouched forest and an uncontaminated agricultural plot. But a realistic look at the earth after a nuclear attack leaves one guessing that a quick, merciful roasting in a personal fireball might be the better way' (Woodwell, 1982, p. 139).

References and Further Reading

Bentley, P.R. (1981) *Blast Overpressure and Fallout Radiation Models for Casualty Assessment and Other Purposes*. London: Home Office Scientific Research and Development Branch.

Bertell, R. (1985) *No Immediate Danger. Prognosis for a Radioactive Earth,* London: The Women's Press.

British Medical Association (1983) *The Medical Effects of Nuclear War*, Chichester: John Wiley and Sons.

Butler, S.J.F. (1981) 'Scientific Advice in Home Defence', in C.F. Barnaby and G.P. Thomas (1982) *The Nuclear Arms Race: Control or Catastrophe*, London: Francis Pinter.

Caldicott, H. (1980) *Nuclear Madness*, New York: Bantam Books Ltd.

Campbell, D. (1980) *New Statesman*, 3 October.

Campbell, D. (1982) *War Plan UK, The Truth About Civil Defence in Britain*, London: Burnett Books.

Clayton, J.K.S. (1978) *Training Manual for Scientific Officers*, Edinburgh: HMSO.

CND (n.d.) *Questions and Answers about Hiroshima and Nagasaki*, London: CND.

Crutzen, P.J. and Birks, J.W. (1982) 'The Atmosphere after a Nuclear War: Twilight at Noon', in Royal Swedish Academy of Sciences (1982) *Nuclear War: The Aftermath*, edited by J. Peterson and D. Hinrichsen, Oxford: Pergamon Press.

DHSS (1979) Guidance Notes for Doctors Teaching Mass Casualty Care.

Ehrlich, P.R. et al. (1983) 'Long-term Biological Consequences of Nuclear War', *Science*, 222, 1293–300.

Glasstone, S. and Dolan, P.J. (1977) 3 edn. *The Effects of Nuclear Weapons*, Washington: US Department of Defense and US Department of Energy, published in England in 1980 by Castle House, Tunbridge Wells.

Greene, O., Percival, I. and Ridge. (1985) *Nuclear Winter: The Evidence and the Risks*, Cambridge: Polity Press.

Greene, O., Rubin, B., Turok, N., Webber, P. and Wilkinson, G. (1982) *London after the Bomb*, Oxford: Oxford University Press.

Home Office, Scientific Advisory Branch (1977) *Training Manual for Scientific Advisers*, London: Home Office.

Home Office and Central Office of Information (1981) *Domestic Nuclear Shelters: Technical Guidance*, London: HMSO.

Home Office and Scottish Home and Health Department (1974 3 edn), *Nuclear Weapons*, London: HMSO.

Laulan, Y. (1982) 'Economic Consequences: Back to the Dark Ages', in Royal Swedish Academy of Sciences (1982) *Nuclear War: The Aftermath*, edited by J. Peterson and D. Hinrichsen, Oxford: Pergamon Press.

Lindop, P.J. (1981) 'Radiation aspects of a nuclear war in Europe', paper presented to the Conference on Nuclear War in Europe, Groningen, Netherlands, April 1981.

Mancuso, T.F. (1977) 'Study of the Lifetime Health and Mortality Experience of Employees of ERDA'. Department of Industrial Environmental Health, Graduate School of Public Health, University of Pittsburg. Mimeograph dated 30 September 1977.

National Academy of Sciences (1975) *Long Term Effects of Multiple Nuclear-Weapon Detonations*, Washington, DC: National Academy Press.

National Research Council of the US National Academy of Sciences, Committee on the Atmospheric Effects of Nuclear Explosions (1985), *The Effects on the Atmosphere of a Major Nuclear Exchange*, Washington, DC: National Academy Press.

Office of Technology Assessment, Congress of the United States of America (1980) *The Effects of Nuclear War*, London: Croom Helm.

Openshaw, S. and Steadman, P. (1982) 'On the Geography of a Worst Case Nuclear Attack on the Population of Britain', *Political Geography Quarterly*, 3, 263–78.

Openshaw, S. and Steadman, P. (1983) 'On the Geography of the Bomb'. Paper presented to the Conference of the Institute of British Geographers, Edinburgh, 5–8 January 1983.

Openshaw, S., Steadman, P. and Greene, O. (1983) *Doomsday. Britain after Nuclear Attack*, Oxford: Basil Blackwell.

Qaswari, A., Wellhoefer, F., and Steward, F. (1982) *Ground Zero*, Milton Keynes: Scientists Against Nuclear Arms.

Royal Swedish Academy of Sciences (1982) *Nuclear War: The Aftermath*, edited by J. Peterson and D. Hinrichsen, Oxford: Pergamon Press.

Sagan, C. (1983) 'Nuclear War and Climatic Catastrophe: Some Policy Implications', *Foreign Affairs*, 62 (2), 257–92.

Sikosa, K. (1984) cited in *The Guardian*, 5 October.

Tame, A. and Rowbotham, F.P.J. (1982) *Maralinga: British A Bomb*

Australian Legacy, Australia: Collins/Fontana Books.

Turco, R.P., Toon, O.B., Ackerman, T.P., Pollack, J.P. and Sagan, C. (1983) 'Nuclear Winter: Global Consequences of Multiple Nuclear Explosions', *Science*, 222, 1283–92.

Woodwell, G.M. (1982) 'The Biotic Effects of Ionizing Radiation', in Royal Swedish Academy of Sciences (1982) *Nuclear War: The Aftermath*, edited by J. Peterson and D. Hinrichsen, Oxford: Pergamon Press.

7
Civil Defence in Britain

Origins

In government and other publications, the terms 'civil defence' and 'home defence' are sometimes used interchangeably. More specifically *civil defence* refers to precautions other than combat taken to protect the civilian population against enemy attack. *Home defence* involves not only mitigation of the effects of an enemy attack but also defence against internal enemies including terrorists, enemy saboteurs and 'dissident extremist groups' who may initiate strikes, anti-war demonstrations and actions disruptive of preparations for war.

Civil defence in Britain had its origins in the experience of bombing in the First World War and the great fear that this engendered in the popular imagination in the 1920s and 1930s. An Air Raids Precautions (ARP) Committee was established in 1924 and an Air Raids Commandant (Designate) appointed in 1933. By 1938 the Air Staff was anticipating that bombing would cause 250,000 casualties in London, together with 3–4 million psychiatric casualties and half the buildings destroyed – all within the first three weeks of war. During the early months of the war the government provided gasmasks, stirrup pumps, and two simple shelters – the 'Anderson' shelter, the vast majority of which were dug into the ground, and the strong steel 'Morrison' table shelter for use indoors. Plans for evacuation were made and about 1½ million people were officially evacuated. Proposals for deep shelters for use by the general public were rejected, although a series of underground bunkers (Citadels) were built for use by key officials in London. Only in November 1941 did the government respond to criticism by announcing that some deep shelters would be built for public use. Overall civil defence proved to be a useful, if sometimes inadequate response, to air raids. At the end of the war the organization was disbanded with the exception of a few units.

Home defence had its origins in the wave of industrial unrest and strikes of 1919–20 and 1926 and the various emergency powers taken by the Government, most notably the Emergency Powers Act (1920) which allows for the proclamation of a state of emergency.

In 1948, in the context of the worsening East–West relations, civil defence was revived. Fear of an atomic war led to the construction of over 100 government and military command and control bunkers, many of which were later to be declared redundant following the development of the H-bomb. A Civil Defence Corps was established and plans for evacuation and dispersal, emergency billeting, food stockpiles, and emergency services were made. By the late 1950s, however, evidence of the fall-out effects of H-Bombs, the development of ballistic missiles, and the upward revision of Home Office estimates of the number of nuclear weapons that might be used on Britain, led the Home Office to revise its plans. What Duncan Campbell (1982) calls 'the spirit of cheerful lunacy' (p. 120) that characterized Home Office publications of the period, such as *The Hydrogen Bomb* (1957) ('Many of the fires caused by a hydrogen bomb could be put out by the methods familiar in the last war, by beating or with a stirrup pump, or with a bucket of sand or water') contrasted strongly with the 1957 Defence White Paper: 'There is at present no means of providing adequate protection for the people of this country against the consequences of attack with nuclear weapons' (Cmnd. 124). Thus 'By 1960, in the official view, civil defence was a means of providing marginal relief to the population, whilst ensuring that the policy of nuclear deterrence remained effective and convincing; the preservation of administration wholly supplanted the provision of relief to survivors as the paramount objective' (Campbell, 1982, p. 120).

Between 1960 and 1963 the Government began to spend less on civil defence. A NATO exercise in 1962 ('Fallex 62') indicated the uselessness of civil defence preparations for large areas of the country. Further reductions in funds followed and finally, in January 1968, 'Home Defence–Civil Defence' was reduced to a care and maintenance basis. The Civil Defence Corps and Auxiliary Fire Service were disbanded, as was, with the exception of a few units, the Home Guard. 'Between 1968 and 1972, and in many ways until much later, civil defence completely disappeared from the public eye' (Campbell, 1982 p. 135). In 1972, however, civil defence re-emerged as home defence, a guise it was to maintain until the early 1980s when the Conservative Government began to establish new civil defence structures.

Home Defence

Objectives and Planning Assumptions 1973–82/3

From 1973 until 1982 the aims of the home defence programme were those given in the Home Office's memorandum ES3/1973:

1 to secure the UK against any internal threat;
2 to mitigate the effects of conventional, nuclear, biological or chemical warfare upon the civil population;
3 to provide alternative government machinery at all levels;
4 to enhance the basis for national recovery in the post-attack period.

It is worth noting that the first objective was the containment of *internal* threats to national security.

Publicly, the Home Office maintained its confidence in civil defence. An undated Home Office leaflet *Civil Defence: Why we need it* (brought out in the early 1980s) categorically rebutted the charge that there is no real protection against a nuclear attack. On the contrary, it asserts that:

> Millions of lives could be saved, by safeguards against radiation especially. But civil defence is not just protection against a nuclear attack. It is protection against *any* sort of attack. NATO experts reckon that any war involving the UK is likely at least to start with non-nuclear weapons.

Critics of these plans argue that the level of destruction caused by a nuclear war, and the nature of its effects on people and the environment, would be such that there would be relatively little that could be done to mitigate the situation short of preventing a nuclear war in the first place. While publicly denying it, this was essentially the view of the Home Office during the 1960s and 1970s. The *Training Manual for Scientific Advisers* (Home Office, 1977) stated that an attack aimed at destroying the UK militarily, politically and economically, and particularly one aimed at destroying dispersed 'retaliatory' forces, would be massive and quick: 'It is likely that the main force of attack by missiles and aircraft would be simultaneous and last no more than 48 hours'. Moreover, 'After a major attack, the United Kingdom could present a scene of enormous destruction and havoc'. Home Office Circular ES10/1974 made it clear that radiation would affect 'a large part of the country', and that considerable areas would have to be abandoned. Circular ES5/1974 bluntly said that 'fire fighting would be undertaken only when the return was judged to be worthwhile and where the survival of organised fire service resources would not be prejudiced'. Circular ES1/1977 stated that 'After a nuclear attack, radioactive fall-out, either in the area or drifting toward it, might be at lethal or near lethal levels. It would be essential that [medical] staff... should not be wasted by allowing them to enter areas of high radioactivity'. Circular ES6/1976 said that 'for planning purposes it should be assumed that in all areas there would be prolonged disruption of piped drinking water supplies.' Circular ES1/1979 said that 'peacetime systems of food processing and distribution would cease to function.' Circular ES8/1976 stated that 'water would not flow from the tap or into the sewerage system. Electricity would be cut off. Refuse collection would cease. Large numbers of casualties would lie where they had died. In such conditions certain diseases could spread rapidly.'

The Home Office circulars to local authorities, from which the above quotations are taken, made it clear that officials were aware of the effects of a nuclear war. In this respect there is little difference between their analysis of the effects of nuclear war on the structure of services and those undertaken by independent observers, although there are, as noted in chapter 6, differences in respect of the estimated number of casualties which might occur. Two points then arise. The first is the very different picture of nuclear war given in the Home Office ES/memoranda from 1973 on, and its statements to the public; and the second is the divergence between the reality of nuclear war and the plans made to cope with it should it ever happen.

The picture portrayed in the Home Office's circulars is so appalling that people have difficulty in accepting it. Indeed, the Home Office seems intent on keeping the public in ignorance of the facts: it has consistently underplayed the effect of a nuclear war in its public statements. Its leaflet *Protect and Survive*, published in 1980 and intended for distribution to all households in the period leading up to an attack, was so incredible that it was received with a mixture of derision, alarm and horror. Far from reassuring people, it discredited the Home Office's public stance on civil defence. The immediate result was an energetic debate about the competence and integrity of the Home Office which is still going on. In 1984, the Home Office was reported as saying 'that it knows nothing about the theory of a nuclear winter'. A Home Office spokesman said 'This is only a hypothesis. It would be foolish to give up any civil defence preparations for something we know nothing about' (*The Guardian*, 6 November 1984). The junior minister responsible for civil defence, Giles Shaw, and his advisers 'rejected an invitation to meet a team of American and Russian scientists including Professor Paul Ehrlich, Dr Richard Turco, Dr Anne Ehrlich and Dr Georgy Golitsyn, on a 10-day tour to explain their findings' (ibid). The overriding impression given by the Home Office's stance has been that it is interested in civil defence only in as much as it provides a cover for home defence.

Before the Attack

Home defence planners make it clear that the transition from peace to war will not be a smooth operation. The extent to which it can be accomplished depends in part on the period of warning given to prepare for an attack.

The Home Office used to assume that there would be from 3 to 4 weeks warning and that the attack would be nuclear. But in 1981 it admitted (ES1/1981) that 'for planning purposes it should be assumed that there may be as little as seven days warning of attack; and the basic essentials should be capable of implementation within 48 hours.'

In September 1980 various government departments conducted an exercise (Operation 'Square-Leg') to evaluate the effects of a nuclear attack on Britain and to test home defence plans. It assumed mounting international tension from the beginning of a given year leading up to an outbreak of war in September. The initial period culminated in August with NATO declaring a state of military vigilance in anticipation of war. The exercise tested the covert civil preparations for war that would have then begun. These entailed the following.

1 The dispersal of key personnel to centres more likely to be protected from an attack.
2 Mobilization of the United Kingdom Warning and Monitoring Organisation (UKWMO). This operates from 873 monitoring posts and is responsible for giving warning of an attack and for monitoring the level of fall-out afterwards.
3 The setting up of the wartime broadcasting service and allocation of police support units to guard key centres.
4 Drawing local authorities into the planning process and instituting various precautions.
5 Dispersal of the fire service into small groups to enhance its survivability.
6 Relocation of Ministry of Agriculture, Fisheries and Food staff to their wartime positions and location and dispersal of supplies of basic commodities.
7 Planning for the implementation of a telephone preference scheme. This in effect closes down the telephone network to all subscribers except those whose access to a telephone is 'vital to the prosecution of war and to national survival after an attack' (ES6/1975). Another group of subscribers who would be able to use their telephones 'in a particular emergency' are those whose lines are 'necessary to maintain the life of the community' (ES6/1975). However, shortly before an attack they would also be cut-off.
8 The health authorities begin discharging patients from hospitals in order to free beds, and then disperse staff and equipment away from cities.
9 The designation of key roads as essential service routes and their closure to the general public by the police, possibly with military assistance.
10 The provision of some public buildings with some protection against blast and fall-out.

Much of the planning was and still is minimal. This is not surprising since the nature and extent of an attack cannot be foreseen. Some local authorities have already earmarked sites for mass burial; others have not. In general, large stocks of disinfectants, lime (for destroying bodies), plastic bags (for putting them in) and so on are not available, yet the presence of bodies will constitute a major environmental health hazard after an attack.

Some of these preparations could be undertaken covertly, but some, by their very nature, would be publicly apparent. At some point the Cabinet would have to approve Queen's Order 2 which suspends Parliament and allows for the assumption of emergency powers. At about the same time the government, through a massive public information campaign, would tell people to prepare for a nuclear attack. However, according to the Home Office (ES2/1975) in the pre-attack period 'Very little material would be released to the public... Government broadcasts might give the first indication of the possibility that war might not be averted, but the emphasis would be on assurances that everything possible was being done to prevent war, and on references to the effectiveness of the nuclear deterrent.' Copies of *Protect and Survive* or its planned replacement, and *Domestic Nuclear Shelters* (Home Office, 1981a), would be distributed. People would be told to lay in a store of non-perishable food sufficient to last them 14 days, and advised to start building shelters. At this stage panic buying is anticipated and supplies would run out. Circular ES1/1979 says: 'Food would be scarce and no arrangements could ensure that every surviving household would have, say, fourteen days supply of food after attack.'

Actual warning of an attack would be a matter of minutes and it is unlikely that everyone would be able to reach shelter before in-coming missiles exploded. Since crucial elements of the early warning system are likely to be early targets, or be damaged by blast and electro-magnetic pulse, and since an attack might go on for several hours, it is also likely that people forced into the open by the effects of earlier bombs would be caught by later ones. Nevertheless, many people might have time to reach some kind of shelter.

Shelters

The Home Office leaflet *Civil Defence: Why we need it* stated that if war threatened, full advice would be given to the public about the warning system and the measures people should take to protect themselves. It was at pains to point out, however, that:

> The risk of war is at present considered so slight that the enormous expense of providing shelters to every family in the land could not be justified. It would cost billions of pounds. As it is, more is being spent on civil defence than previously – about £45 million a year by 1983/84. This is an 'insurance premium' against the remote risk that NATO's continuing deterrent policy might fail.

The leaflet claimed that most senior ministers, government officials and service chiefs would 'take their chance like anybody else' in an attack. It denied that any special provision was being made for them but pointed out that there were plans for government to be dispersed to civil defence regions. Decentralized war headquarters would be activated, but 'although reinforced', these 'are certainly not invulnerable'. Indeed, most civil defence operations in a war emergency would be directed by local officials working in the basements of town halls and similar premises. While government bunkers are still, in general, far below American standards, they do exist, and there are plans to evacuate leading members of the community. As David Owen, the leader of the Social-Democratic Party, has commented, under Home Office plans 'the governers will go underground, the governed will stay on top' (Campbell, 1982, p. 156).

So far as individuals are concerned, the emphasis in *Protect and Survive* was on the calm preparation of either a fall-out room and, within that room, an inner shelter made of tables and heavy objects which would shield the inhabitants from the effects of fall-out, or a home-made garden shelter. The government would not provide shelters because, as Lord Belstead, then in charge of civil defence affairs at the Home Office, said in a 1980 television programme, the nation cannot afford it. As the Home Office Minister responsible for civil defence, Douglas Hurd (1983) said, given the unlikelihood of nuclear attack, 'the Government would hardly be justified in diverting thousands of millions of pounds of scarce resources into, for example, a national shelter building programme.'

Of course, such a programme would take years to provide. It would be too late to begin one in the months or weeks of a deteriorating international situation. The government has, however, provided technical advice on the building of shelters in two Home Office booklets, *Advice on Domestic Shelters providing protection against nuclear explosions* (1981a) and *Domestic Nuc-*

lear Shelters: Technical Guidance (1981b). Two of the designs mentioned in the former were based on trench and field shelters of great discomfort that offer no blast protection (the *Architects Journal* dismissed these designs 'out of hand' (Campbell, 1982, p. 455)); two were based on the Second World War Anderson and Morrison shelters; one was an expensive underground concrete type. The accompanying *Technical Guidance* failed to say how much radiation would be lethal, making any assessment of the efficacy of the shelters' PF factor impossible.

For a short while (January to June 1981) a national magazine, *Protect and Survive Monthly*, tried to capitalize on advertisements for nuclear shelters. Private enterprise stepped in, with several hundred manufacturers and builders offering shelters, many of dubious or even no value.

Since the rudimentary shelters proposed by the government will not withstand blast pressures, the failure to provide any form of deep shelters could be regarded as a callous failure to protect that element of the population at risk from the effects of blast. Those who press for a nuclear-shelter building programme argue that, faced by a likely nuclear war, it is the government's prime responsibility to protect its people. Doris Lessing (1982) protested that in the face of this threat, 'governments have conspired to leave us helpless.' Perhaps for the first time in history 'governments have reneged on their main responsibility'. She had no time for those who regard civil defence as useless. The failure of citizens to inform themselves about the reality of threats facing the country and to require the government to do something about it is, she suggested, even more surprising than the government's failure to act. Impressed by the mass sheltering programmes in Switzerland and Sweden, she asked why large numbers of people reject a nuclear-shelter building programme and concluded that 'we may be witnessing a protective psychological mechanism that enables the organism to accept, slowly, at a safe pace, the fact that it really does have to adapt to prepare for possible war.' Nobody wants to believe that a war might come – hence the rage which greets those proposing the building of shelters.

Opponents of civil defence agree with the government that the cost of providing adequate sheltering for everyone would be astronomical. Since the size and pattern of a nuclear attack cannot be reliably predicted, a comprehensive sheltering programme would have to provide blast-proof sheltering for the entire population. Shelters in areas where firestorms might occur would need independent air supplies. There is no way in which such a programme could be provided. It is economically and therefore politically impracticable. But, even supposing that deep shelters could be provided for the whole population, what kind of a world would people emerge into, and how would they survive in a world devastated by nuclear weapons? The evidence on the long-term effects of a nuclear war suggest that the provision of deep shelters could not protect the population from these effects.

It is also suggested that the government could not provide nuclear shelters without publicly admitting that the policy of deterrence is flawed. If it provides shelters either it admits the likelihood of a nuclear war, calls in question the whole concept of deterrence, and opens the way for its nuclear defence policy to be questioned; or it is open to the charge that it is planning to fight a nuclear war while making some prior provision for its citizens. If it does not provide shelters, it risks being accused of callous disregard for the safety of its people. In effect, the policy of requesting large numbers of people to take shelter in their own homes and in make-shift shelters condemns them to death. If a deep-shelter programme is too costly and politically embarrassing, an alternative policy might be the large-scale evacuation of the population from likely target areas.

Evacuation

Until 1983 plans were based on the assumption that the civilian population would remain at home in the period preceding an attack. *Protect and Survive* said that the reason for this policy was that 'No part of the United Kingdom can be considered safe from both the direct effects of the weapons and the resultant fall-out.' Indeed, the government positively discouraged any movement of the population. On the other hand,

> certain built up areas may be regarded as potential targets... If the total destruction or isolation of health service resources is to be avoided, some redeployment of medical and nursing staff, medical supplies, ambulances and equipment would be essential. The major concentration of hospitals lies in the centres of large cities and

> towns... The redeployment of resources could reduce the possibility of total destruction (Home Office, ES1/1977).

As Duncan Campbell comments (1982, p. 375) 'this is perhaps as close as the government has come (in an unclassified document) to acknowledging the writing off of the cities as an act of policy.' It is also an admission that civil defence has been aimed at mitigating the effects of a *nuclear* attack, since preparations for conventional attacks would argue for a degree of concentration of resources in reasonable proximity to urban centres.

This position has been challenged by Duncan Campbell (1982). He says that Home Office experts privately assumed that while 65 per cent of the population would survive a nuclear attack if they stayed put, 87 per cent would survive if they dispersed (1982, p. 403). The official policy of controlling access to major routes leading out of the cities in time of war such that, as *The Guardian* (23 November 1982) put it, 'only people who would be needed for specific functions after hostilities ceased – including doctors – would be allowed to leave their homes', would by Home Office estimates have resulted in an additional 12 million deaths. Moreover, although the leaflet *Civil Defence: Why we need it* gave a picture of government ministers, senior civil servants, and service chiefs sitting at their desks up to the last moment, the North East Thames Regional Health Authority war plan, which was leaked to the press, noted the need to secure the survival of the nation's leaders and those people who may be useful in recovery after an attack.

In 1983, however, *The Sunday Times* (30 January) reported that Home Office officials were considering 'radical proposals for the evacuation of civilians in time of war.' Under one option, 12 million people would be moved from the big cities and other likely targets, while another, even more ambitious proposal envisaged resettling almost two-thirds of the population. However, a report in *The Guardian* (14 May 1983) pointed out that the Government's draft legislation on civil defence had dropped any mention of evacuation, but this was followed by a further report (*The Guardian*, 11 July 1983) to the effect that the Government was considering the evacuation of those living near US and British nuclear bases.

It is difficult to envisage how any evacuation scheme might work, particularly in a relatively small, heavily populated country such as Britain. Plans to evacuate the population are usually criticized, firstly, because they might signal preparations for a first-strike and hence *precipitate* a pre-emptive strike by the enemy; secondly, they assume that the population will act in a calm and rational manner and that the evacuation could be carried out smoothly; thirdly, the evacuees would still need to be provided with shelter against the effects of fall-out; fourthly, plans would have to be made to provide food and water in the immediate aftermath of the attack; and fifthly, the longer term problems of survival (food, water, contamination, and the problems of a nuclear winter) remain.

After the Attack

The fact that civil defence measures are unlikely to protect the population during an attack does not necessarily negate their usefulness after one has taken place. On the other hand, the government has no illusions about the likely conditions after a nuclear war: 'Each side would have the capacity to inflict upon the other a degree of devastation which has never before in human history been either possible or imaginable. An armed clash involving the interests of either side is, therefore, likely to lead to the virtual destruction of both and not merely to conquest or defeat' (Ministry of Education, 3/1964).

The United Kingdom Warning and Monitoring Organisation (UKWMO) is intended to monitor fall-out and provide information on this to group and sector controls. Whether it would be able to do the latter is doubtful. A large nuclear explosion in the atmosphere disrupts radio communication and, by creating electric fields at ground level, can permanently damage the sensitive components in transmitting and receiving equipment. The temporary disturbance may last minutes or hours, but the damage to equipment caused by the electromagnetic pulse (EMP) induces high voltages in any aerial systems at ground level. On long power-transmission lines the induced voltages may be so high that it shuts down the whole grid system. It would also destroy the sensitive solid-state components in all modern radios and telecommunications equipment and in computers. This effect is produced in an area equal to the horizon as seen from the point of the explosion. An explosion at a height of 150 miles would therefore produce an EMP effect in an area equal to a circle of 2,000 miles diameter. This would be sufficient to black out most of the continent of Europe. The UKWMO relies heavily upon communications run through land lines. Since these are likely to be damaged by the EMP, it is unlikely that its communication system would survive a nuclear attack. According to Duncan Campbell (1982, p. 323). RAF bunkers are shielded against EMP.

Government plans have emphasized that what is left of the various services should be used cautiously after an attack. The problem of radiation means that the fire, rescue and medical services will only begin to operate when it is judged that it is safe for them to do so. Plans for the billeting of homeless people envisage a housing density of one family per room. The repair of essential services such as water, gas, electricity and sewage systems and bridges and houses, the clearing of roads, and the construction of communal toilets, will be a high priority. The construction industry will be brought under the control of the Emergency Works Organisation. Emergency feeding will begin as soon as possible. However, many of the plans for feeding survivors seem quite unrealistic, for example the plan for Bexley in South London talks about providing special meals to be taken to the infirm, injured and seriously ill, emergency week-long training courses for food-centre supervisors and typical menus of stew, potatoes, greens and tea. The plan contradicts itself: 'the fuel normally used in meals kitchens will be gas,' yet 'no reliance can be placed on the availability of gas' (Campbell, 1982, p. 355).

The distribution of food is, in fact, likely to be a major problem. Gas pipes would be damaged. The four points at which natural gas pipelines from the North Sea come ashore may be targets and, if destroyed, the UK's natural gas supply network would cease to operate. So far as the coal industry is concerned, surface installations may be destroyed by blast, while the failure of the electricity system would cause pumping to cease and lead to the flooding and collapse of some pits. Oil imports would probably cease. The destruction of refineries and off-shore installations would lead to the almost total loss of domestic production. The Home Office's conclusions are that 'For planning purposes it may be assumed that, after a nuclear attack, all energy production and supply would soon cease' (Home Office, ES5/1976).

Burial of the dead will pose an immense problem, as will the disposal of human waste. Medical services will be overwhelmed. Supplies of drugs will quickly be exhausted and it may be a very long time after the attack before they can be replaced (Health Department, 1978). Treatment will be offered only to those individuals who have a chance of survival. The loss of power and shortages of drugs and blood for transfusions will severely limit the level of medical services offered. The term 'cottage medicine' is frequently used in the plans. Campbell (1982, p. 367) suggests that as far as the Health Service goes 'The war planning effort against a significant background of internal resistance within the National Health Service is perhaps the most cynical of all the civil defence preparations. No money is spent on stockpiles of special facilities.' Indeed, the Health Service cuts have reduced provision, and 'palliative paper plans will save no one's suffering or death' (p. 368). Casualties would be sorted into three categories. Firstly, those who will survive with no further treatment: they are sent home or to rest centres. Secondly, those who with limited medical help might be alive after 7 days and have a fair chance of eventual recovery: they would be admitted to Casualty Collecting Centres, where conditions are likely to be horrific. Thirdly, those suffering from major injuries, who would be sent to 'holding units' and left to die. At the Casualty Collecting Centres, triage (the process of sorting casualties into these three categories) would be continuous. Radiation casualties would not be treated at all. Those likely to die will be denied food and may well be shot to put them out of their misery. 'Unproductive' members of society (the old or the very young, the sick or mentally retarded), are also likely to receive no medical help.

Use of the telephone system will continue to be restricted to priority subscribers (Home Office, ES5/1975) and similar restrictions will operate in respect of letters. The production of energy will be significantly reduced and may be brought to a halt. The national electricity grid system will have been damaged by blast and EMP. Large conventional and all nuclear-power stations are likely targets for an attack. If a nuclear reactor is hit by a bomb, the radioactivity contained in it will be dispersed. Since nuclear reactors contain a high proportion of long-lived radioactive material, the long-term environmental effects of the attack are horrifying. The Royal Swedish Academy of Sciences (1982) study concluded that an attack on nuclear power stations would render the territories of the countries attacked, and neighbouring countries 'uninhabitable by ordinary standards of radiation safety for years or decades' (p. 46). If the attack were directed against reprocessing plants or waste storage tanks, the uninhabitable areas would be even greater and they would remain so for an even longer period. Such attacks would have serious consequences for civil defence planning.

Continuity of Government

Up to 1983, Home Office plans envisaged the structure of government shown in figure 7.1. This provided a means by which government would continue to exist at some levels – probably at the sub-regional or county level. It was heavily criticized on the grounds that all forms of democratic control would be swept away. The Home Office placed considerable emphasis on the sub-regional level. Its circular ES7/1973 suggested that at least initially the sub-regional headquarters would be the highest effective level of internal government. Although civilian regional governments were to be established 'as soon as practicable' there are still no

Figure 7.1 The structure of government in a post-war Britain as envisaged in 1973

National seat of government

|

12 Regional seats of government

|

23 Sub-regional headquarters

|

County controls

|

District controls

|

Sub-district controls

|

Community posts of rest centres

Note: Latest plans show 11 regional seats of government, 9 of them divided into two zones. The sub-regional tier has been abolished.

bunkers for them. However, Greene et al. (1982, pp. 82–3) pointed out that there was one Armed Forces Headquarter per region, in a deep bunker containing the regional military commander and his staff. They drew attention to the war scenario described by General Sir John Hackett (former Deputy Chief of the General Staff and Commander of the Northern Army Group of NATO) and others (1978) where it was said that after the war 'delegation of power over military *and civil defence* resources had become the responsibility of the Headquarters of the UK Land Forces' (p. 389) (my emphasis). This suggests that far from taking a relatively secondary role in support of the civil powers it was envisaged that the military would provide the higher 'effective level' of government.

Many critics of home defence argue that its chief purpose is to maintain the government and to ensure that any opposition is crushed. Repressive measures in the lead-up to a nuclear attack include the arrest of 'potential subversives' and the control of crowds attempting to leave the cities. After an attack, maintenance of law and order would, in the 1973 plans, have been a responsibility of the sub-regional commissioner. Justice would be administered by 'those holding any form of judicial office and all active justices of the peace'. They would be able to hold courts anywhere in their region. In their absence, lay commissioners would be able to hold court, either alone, in pairs, or as a bench of three. The right to trial by jury would be largely abolished although 'in capital cases, wherever practicable, there would be a jury of not more than five, empowered summarily, or a court consisting of not less than three commissioners'. There would be no appeals, although the senior commissioner would make arrangements to review the decisions of emergency courts in his area. The Home Office envisaged penalties such as communal labour, restricted rations and exposure to public disapproval as possibly being appropriate for all but the gravest offences, but 'in the case of flagrantly anti-social behaviour there might be a need for harsher penalties' including punishments 'not normally available to the courts'. Most critics assume this referred to summary executions (Home Office, 1976, cited in Bolsover, 1982, p. 37).

The West Sussex County Council's Community Survival Guide (1977) speaks of restrictions on potentially subversive people, and suggests that the homeless 'should be collected into groups only for the shortest possible period and that every effort should be made to avoid large groups' in order to 'lesson the danger of epidemics, to keep up morale *and to avoid the ingredients of law and order problems*' (my emphasis). Local peace-keeping forces would be established. Since detention of offenders 'may not be possible', 'punishments will probably be corporal and immediate'. Each peace-keeping force would have a 'cadre of Street Watchers'. Some documents speak of 'internment areas' (cited in Bolsover, 1982, p. 50).

At the heart of this concern is a fear that law and order will collapse and that the people will wreak vengeance on the government that brought disaster upon them. In 'defending' democracy from an external threat, the government will be forced to rigorously suppress it at home after a nuclear war. Substitute for the phrases and words used in the official documents some more common descriptions and a frightening picture of society after the bomb emerges: summary arrest, concentration camps, summary trial and execution, forced labour and local informers are all envisaged.

The Revival of Civil Defence 1983/4

Since the early 1980s the Home Office has tried to reactivate both civil defence voluntary groups and local government plans but, given the unreality of civil defence planning, many local authorities have refused to cooperate. This led to the cancellation of the Home Office civil defence planning exercise Operation 'Hard Rock', due to have been held in late September and early October 1982. In response the Government published draft regulations in March 1983 which sought to force local authorities to establish, equip and maintain wartime emergency centres, train council and district staff as well as volunteer civil defence forces, and locate and use suitable buildings, structures and excavations as public shelters. One of the effects of the June 1983 General Election was to postpone the passage of the government's new Civil Defence Regulations. However, following their victory, the Conservative Government laid the Draft Regulations before Parliament on 12 July 1983, and on 1 December they came into force (Civil Defence (General Local Authority Functions) Regulations, 1983).

Critics complained that the original draft regulations were ambiguous. They pointed out that it was impossible to make local plans and prepare for a nuclear war and its aftermath without information about the possible nature of the attack for which one was planning. Would it be a conventional or a nuclear attack? How big would it be? Which targets would be hit? The Home Office has refused to provide answers. Simon Turney, the Greater London Council's councillor with responsibility for the rescue services, pointed out that 'It's very hard to know what the legislation really means. The whole draft is full of this bloody weazel word 'suitable'. What is a suitable shelter?... What is a suitable volunteer?' (Turney, 1983). The regulations were opposed by 155 councils which had declared themselves to be 'nuclear free'. A survey by *Local Government News* revealed that one-third of chief executives and two-thirds of all other officers would not take up their war-time emergency roles in the event of a war. Even the largely Conservative Association of County Councils opposed the regulations because it was 'worried about money and hours' (*The Guardian*, 14 May 1983). The funds available – £56 million, or about £1 per head of the population per year – are wholly inadequate to build reinforced air-filtered control rooms, let alone stock up on essential supplies.

The regulations require councils to

1 establish, equip and maintain two (or in the case of Greater London, five) emergency control centres (most recently redesignated 'emergency centres');
2 train appropriate numbers of staff and recruit, organize and train networks of volunteers in civil defence duties;
3 take part in civil defence exercises;
4 plan to use buildings, structures, excavations and other suitable features of land for public civil defence shelters;
5 plan to provide and maintain a rescue service for people; and
6 plan to secure the help of voluntary organizations and individual volunteers for civil defence work.

Over 150 local authorities in Britain have declared themselves nuclear-free zones. Such councils have, in general, accepted a commitment to providing people with information about the dangers of nuclear war; opposed the siting or manufacture of nuclear weapons within their boundaries and the transport or storage of nuclear waste through and in their area; and questioned the value of the government's civil defence plans. While a few of these local authorities have refused outright to implement the regulations and some have decided to comply with the absolute minimum, the majority have called upon the Home Office to clarify the assumptions about the nature and scale of a nuclear attack for which they are supposed to plan, arguing that it is unreasonable to plan, or even locate, emergency control centres in the absence of such information. After a great deal of delay, the Home Office finally issued new guidance in 1984 in the form of three circulars, ES1/1984 *The Government's Planning Assumptions for Civil Defence*; ES2/1984 *The Revised Arrangements for Wartime Machinery of Government*; and FS6/1984 *War Emergency Planning for the Fire Service*. It also issued a draft version of revised and consolidated guidance, which it sent to the local authority association for its comments.

The *Consolidated Circular to Local Authorities on Emergency Planning* issued by the Home Office (1984) provides information for the guidance of local authorities in the implementation of the new regulations. The emergency centres are expected to have a protection factor (PF) of 100 and be able to withstand a blast over-pressure of 1.5 psi. They should be able to operate for 14 days without mains electricity and should be capable of being made fully operational within 48 hours. However, while the Home Office will make grants available to cover up to 75 per cent of the cost of improvements, 'in the present circumstances approvals are likely to be limited to schemes costing not more than £30,000.'

On the question of staff training, local authorities are compelled to train an 'appropriate' number of 'suitable' staff in civil defence duties. However, while

> thorough peacetime briefing and training will be necessary in the case of those officers designated for key roles in directing the wartime implementation... less intensive briefing and training may be considered adequate in the case of staff at lower levels of responsibility while, for the majority of the authority's staff it would suffice for there to exist a carefully planned programme for crash briefing and training in a war emergency.'

Such courses, provided in the period of tension, would involve 'simple written and oral instruction'.

Following the cancellation of the Hard Rock civil defence exercise in 1982, there are no immediate plans for a national exercise of this kind. Although all local authorities have a duty to take part in such exercises, the *Consolidated Circular* only says that 'Local authorities are encouraged to commence a programme of local exercises as soon as they are satisfied that the preparations of plans and progress on training is sufficiently advanced to permit a successful exercise to take place.' Exercises at district or county level are proposed, with, subsequently, possible regional exercises. 'The HO [Home Office] will revise the scope for more extensive exercises in further years when the civil defence preparations are more advanced'.

ES1/1984 suggests that planning needs to be related to four stages: firstly, a period of international tension; secondly, conventional attack with the possible use of chemical weapons; thirdly, nuclear attack; and fourthly, survival and recovery after a nuclear attack. The circular suggests that a conventional attack on Britain is more likely than an initial nuclear attack and that, moreover, such an attack might continue for several weeks. Indeed, the 'main threat to the UK would be from conventional air attack' which would probably be 'confined to targets of military significance'. Neither the size nor the targets of a conventional or nuclear attack can be predicted. The circular states: 'It is assumed that... air bursts of up to five megatons might be used to destroy city areas... In some instances more than one weapon might be used to ensure destruction of a target.' In a major attack on cities the numbers of killed or injured could 'amount to many millions' and industries, services and communications could be destroyed 'possibly beyond reconstruction in some areas'. After an attack on military installations 'everywhere there could be a danger from radioactive fallout'. Moreover, 'areas of intense radiation could persist for months'. Generally, while many millions of people might be left alive, even after a major attack on cities, 'no clear assumptions can be made about the survival or recovery of the country... following a nuclear attack.'

The suggestion that a conventional attack is the most likely initial form of attack represents a major change in Home Office assumptions. ES3/1973 stated that 'The current assessments point clearly to nuclear war as the overriding consideration in determining preparations to be made for Home Defence in the UK.' Moreover, while the Home Office now says that it cannot predict the scale of the attack nor the targets, previous Home Office scenarios (with the exception of the cancelled Hard Rock exercise) envisaged an attack of about 200 megatons. Knowledge of Soviet targeting policy and of potential targets in Britain allows some degree of certainty in predicting the likely pattern of an attack (see, for example, Campbell, 1982, pp. 424–6; and Openshaw et al. 1983, pp. 242–69). It is also likely, given the integration of nuclear and conventional weapons on the Central Front, that a war between NATO and Warsaw Pact forces would quickly 'go nuclear'. Once this happened, Britain could expect to be subjected to a nuclear attack as the Soviet Union tried to destroy British and US nuclear forces based in the UK. By then it would be highly unlikely that the Soviet use of nuclear weapons would be restricted to military nuclear forces.

As international tension mounted, Regional Emergency Committees would be activated 'to prepare the nation for war' and to 'co-ordinate civil and military requirements'. Following a nuclear attack, the powers of central government would devolve to a Regional Commissioner – a government minister who would exercise full emergency powers within each of the 11 home defence regions (one less than in the previous structure). Nine of the eleven regions would be divided into two zones, one with the Regional Commissioner's headquarters, the other with a second headquarters under the command of a Deputy Regional Commissioner. The sub-regional tier (see figure 7.1) has been dropped.

Circular FS6/1984 departs from previous advice in making no plans for the dispersal of fire-engines and crews from cities during periods of tension. This is consistent with the emphasis on conventional attack. Brigades are expected to identify 'havens' where their fire crews can go for protection if they are caught in the open during a nuclear attack.

The *Consolidated Circular* document replaces the series of ES guidance circulars issued to local authorities between 1971 and 1982, and provides guidance on the actions which local authorities should take to implement the new civil defence regulations. Interestingly, the unequivocal advice to the public to stay put in their homes has disappeared. While 'the best protection for the public is to stay at home using and improving the protection of their homes,' this policy 'is purely advisory'. There is no direct statement on the closure of Essential Service Routes, although 'traffic management would be the responsibility of the police.' The police Manual of Home Defence remains more explicit: 'Priority should be given to the Essential Service Routes which, so far as possible, should be kept clear of refugees and non-essential traffic.'

The direct reference to securing Britain against internal threat contained in ES3/1973 has gone. Section 21 of the *Consolidated Circular* explains that one of the army's tasks in Home Defence is to defend Britain against 'any enemy threat'. The Police Manual of Home Defence is less circumspect. Police war duties include 'Special measures to maintain internal security with particular reference to the detention or restriction of movement of subversives or potentially subversive people.'

Conclusions

There is widespread doubt about the effectiveness of British civil defence measures in the face of the nuclear threat. In 1957 a Conservative Government declared: 'There are no means of providing adequate protection for the people of this country against the consequences of an attack with nuclear weapons' (Cmnd. 124). In the 1980s, the situation is even worse. The number of weapons targeted on the West (including Britain), and their accuracy, has increased. Nevertheless, it is clear that a misguided, and misguiding, belief in civil defence is held by members of the British government. Leon Brittan, then Minister of State at the Home Office, was reported as saying that 'Civil preparedness should be adequate if the credibility of the military deterrent strategy [is] to be maintained. Military and civil preparedness [are] closely related' (Brittan, 1980). The most dangerous illusion is that civil defence will indeed provide a significant degree of protection against the immediate and long-term effects of nuclear war, and that people will come to believe that survival is possible.

References and Further Reading

Bolsover, P. (1982) 2 edn. *Civil Defence: the Cruellest Confidence Trick*, London: Campaign for Nuclear Disarmament.

British Medical Association (Board of Science and Education) (1983) *The Medical Effects of Nuclear War*, Chichester: John Wiley & Sons.

Brittan, L. (1980), cited in *The Times*, 23 February.

Campbell, D. (1982) *War Plan UK. The Truth about Civil Defence in Britain*, London: Burnett Books.

Clayton, J.K.S. (1978) *Training Manual for Scientific Officers*, Edinburgh: HMSO.

Cmnd. 124 (1957) *Defence: Outline of Future Policy*, London: HMSO.

Crossley, G. (1982) *Civil Defence in Britain*, Bradford: University of Bradford School of Peace Studies/London: Housmans.

Draft Civil Defence (General Local Authority Functions) Regulations 1983. Laid before Parliament, 12 July 1983.

Greene, O., Rubin, B., Turok, N., Webber, P. and Wilkinson, G. (1982) *London after the Bomb*, Oxford: Oxford University Press.

Hackett, General Sir John, and others (1978) *The Third World War: August 1985*, London: Sidgwick and Jackson.

Health Department (1978), Circular HDC (78) 1, *Medical Supplies in War*.

Home Office (n.d.) *Civil Defence: Why we need it*, London: Central Office of Information.

Home Office (1957) *The Hydrogen Bomb*, London: HMSO.

Home Office (1976) Briefing Material for Wartime Controllers.

Home Office, Scientific Advisory Branch (1977) *Training Manual for Scientific Advisers*, London: Home Office.

Home Office (1980) *Protect and Survive*, London: HMSO.

Home Office (1981a) *Advice on Domestic Shelters: providing protection against nuclear explosions*, London: HMSO.

Home Office (1981b) *Domestic Nuclear Shelters: Technical Guidance*, London: HMSO.

Home Office (1984) *Consolidated Circular to Local Authorities on Emergency Planning*.

Home Office, ES memoranda, 1973 to date.

ES3/1973, *Home Defence Planning Assumptions*.

ES7/1973, *Machinery of Government in War*.

ES5/1974, *War Emergency Planning for the Fire Service*.

ES10/1974, *Public Survival under Fall-out Conditions*.

ES6/1975, *Post Office Telephone Preference Scheme*.

ES5/1976, *Energy Supplies in War*.

ES6/1976, *Water Services in War*.

ES8/1976, *Environmental Health in War*.

ES1/1977, *The Preparation and Organization of the Health Service for War*.

ES1/1979, *Food and Agricultural Controls in War*.

ES1/1981, *Home Defence Review*.
ES1/1984, *The Government's Planning Assumptions for Civil Defence*.
ES2/1984, *The Revised Arrangements for Wartime Machinery of Government*.
FS6/1984, *War Emergency Planning for the Fire Service*.
Hurd, D. (1983) cited in *The Guardian*, 23 September.
Laurie, P. (1980) *Beneath the City Streets*, London: Panther.
Lessing, D. (1982) *The Guardian*, 25 September.
Ministry of Education (1964) Circular 3/1964, *Education Service in War*.
Openshaw, S., Steadman, P. and Greene, O. (1983) *Doomsday. Britain after Nuclear Attack*, Oxford: Basil Blackwell.
Owen, D. (1980) *Negotiate and Survive*, cited in Campbell, D. (1982) *War Plan UK. The Truth about Civil Defence in Britain*. London: Burnett Books.
Rogers, P., Dando, M, and van den Dungen, P. (1981) *As Lambs to the Slaughter. The Facts about Nuclear War*, London: Arrow Books.
Royal Swedish Academy of Sciences (1982) *Nuclear War: the Aftermath*, edited by J. Peterson and D. Hinrichsen, Oxford: Pergamon Press.
Turney, S. (1983) cited in *The Guardian*, 14 May.
West Sussex County Council (1977) Community Survival Guide.

8
The Legal Status of Nuclear Weapons

Since the middle of the nineteenth century there has been a growing corpus of international law, variously known as the international law of armed conflict or the humanitarian laws of war, which has major implications for the legality of nuclear weapons.

Although the sanctions of international law are tenuous people are entitled to demand that their leaders will obey the law, not least because as Blackstone put it in his *Commentaries* on English Law, 'the Law of Nations is part of the Law of the Land'. This judgment was confirmed by the British Prosecutor at the Nuremberg Military Tribunal, Sir Hartley Shawcross, who pointed out that 'In England and the United States our Courts have invariably acted on the view that the accepted customary rules of the Law of Nations are binding upon the subject and the citizen, and the position is essentially the same in most countries' (*International Military Tribunal*). He went on 'If international law is to be applied at all, it must be superior to State Law in this respect, that it must consider the legality of what is done by international and not by State law tests.'

Many individuals opposed to nuclear weapons are recognizing that international law provides a vehicle for their protest, and there are growing attempts to make governments comply with international treaties. Some have also argued that information on the implications of international law for nuclear weapons should be brought to the attention of personnel in the armed forces. This was recently done in Britain by Monsignor Bruce Kent, the General Secretary of the Campaign for Nuclear Disarmament, in an article in *The Guardian*. He focused on the deployment of US cruise missiles in Britain and urged servicemen who were responsible for the security of these missiles '(a) to request duty not connected with cruise missiles, (b) if that is not possible to make arrangements to leave the Armed Forces, (c) if no other course of action is possible to refuse to obey illegal and immoral cruise related orders' (Kent, 1983). In a linked article, *The Guardian*'s reporter Paul Brown pointed out that The Incitement to Disaffection Act (1934) makes it an offence to try to seduce members of the British armed forces from their duty or allegiance. Another view of the obligations of individuals is, however, provided by Professor Richard Falk, who pointed out in a speech at Nuremberg in 1983 that 'Everyone has the right and duty to say No to illegal State policy... It is not disobedience but the enforcement of law to refuse to be an accomplice to the preparation of nuclear war'.

Historical Background

Just-War Theory

From earliest times there have been attempts to define the grounds on which war may be declared and to control in some degree the conduct of hostilities. From the fourth to the seventeenth century, lawyers were much preoccupied with the theory of the just war. This was defined as one which was not only fought for a just cause, but was conducted in a manner which was not condemned by international law (or the 'law of nations' as it was then called). Developed out of Greek, Roman, Jewish and Christian roots, just-war theory reached its highest development during the sixteenth and seventeenth centuries. Taking as its starting point an acceptance of war as a fact of life, it laid down criteria by which a war could be judged to be fought in a just cause and manner.

The theory holds that war can only be entered into by a formal declaration of war issued by the sovereign power and that it must be fought only as a last resort for a just cause, of which historically there were three: firstly, to punish an offence; secondly, to regain territory or goods seized by an aggressor; and thirdly, to repel aggression. Only the third of these is now generally accepted with the exception, of course, of pacifists. It was also argued that it was not sufficient for one's cause to be just. Those engaged in a war had to have a reasonable hope of winning. Moreover, the evil and damage caused by the war had to be judged to be proportionate to the injury it was designed to avert or the injustice that occasioned it. Justness of war was also determined by the justness of the means used to wage it. In any war, the rights and safety of non-combatants were to be respected. This did not necessarily mean that the casual and unintentional killing or injury of non-combatants was wrong, but that care should be taken not to employ means that cause unnecessary danger to such people. Lawyers also argued that the conduct of war must not result in a disproportionate amount of damage. The intention should not be to ruin the enemy nation but to protect rights and rebuild peace and order in the international community.

At the time the theory was being developed wars were, mainly, fought by small professional armed forces. The modern concept of total war, in which a whole society is deemed to be at war and in which the distinction between military and civilian objects and personnel is blurred, had not been considered.

Interest in just-war theory waned during the eighteenth and nineteenth centuries as it came to be recognized that there was no way in which sovereign states could be brought to account for the justice of their cause. War was increasingly treated as an activity which began when a nation initiated a hostile act. The practice of issuing a formal declaration of war, which had been required under just-war theory, was abandoned. International lawyers sought instead to define the rights and duties of neutrals and focused their attention on the relations between states at peace with each other and the avoidance of war through diplomatic intercourse, arbitration and mediation. Many of the principles of just-war theory have, nevertheless, been incorporated into the modern international law of armed conflict – particularly the protection afforded to non-combatants and the principle of proportionality – and they are therefore discussed below.

The Development of International Laws governing War

Continued experience of war between European nations gave an impetus to the development, during the latter half of the nineteenth century, of international laws controlling the conduct of war. A series of meetings at Geneva in 1863–4 led to the establishment of the Red Cross and the drawing up of the first code for the care of sick and wounded soldiers. The Declaration of St Petersburg (1868) prohibited the use of explosive bullets. The Brussels Congress of 1874 drew up elaborate rules for the conduct of war which, although never implemented, paved the way for the conferences at The Hague in 1899 and 1907. These concentrated upon rules of war for land and sea, and prohibited certain weapons (dum-dum or expanding bullets, and the use of poison or poisoned weapons). The same period saw national governments follow the lead of the USA in drawing up manuals on the conduct of war for use by their field commanders.

The revulsion against war which followed the First World War led to a rebirth of the idea that, except in approved circumstances, war is unacceptable and unlawful. The Covenant of the League of Nations (1920), the Paris Pact of 1928 and the Argentine anti-war pact of 1933 reinforced these views. Although attention during the inter-war years was focused on the prevention of war rather than on its conduct, there were a number of attempts to prohibit the use of particular weapons, notably those curbing submarine warfare initiated by the 1922 Washington Disarmament Conference and controlling air bombardment through the drafting of The Hague air rules (1923), but these initiatives failed.

The experience of the First World War, in which the extensive use of poisonous gas resulted in as many as 1,300,000 casualties, led to a horror of gas. Article 171 of the Treaty of Versailles specifically prohibited 'the use of asphyxiating, poisonous or other gases and all analogous liquids, materials or devices'. However, the Treaty was applicable only to the vanquished countries (Germany, Austria, Hungary and Bulgaria). As a result the Geneva Protocol (1925) was signed, prohibiting the use of asphyxiating, poisonous or other gases, and of all anologous liquids, materials or devices, as well as the use of bacteriological methods of warfare. Between 1925 and 1939 over 40 states subscribed to the protocol. Although Italy violated this convention during its war against Ethiopia in 1935–6, the states engaged in the Second World War observed the conventions against gas and also forbore to use bacteriological weapons. Apart from the Gas Protocol, however, none of the other international agreements proved sufficient in providing humanitarian safeguards during the Second World War.

In the aftermath of the War, the Convention on the prevention and punishment of the crime of genocide (the 1948 Genocide

Convention) was agreed. Further rules on the conduct of war were then worked out at a conference held in Geneva in 1949 and included in four Conventions primarily concerned with laying down a code of behaviour in wars of the traditional type, conducted between states and between regular armies (the Geneva Conventions). They have, however, proved to be imperfect instruments for controlling behaviour in guerrilla wars and, as a result, the degree of protection afforded civilians has been weakened. To deal with some of these problems, a diplomatic conference on the reaffirmation and development of international law applicable in armed conflicts was convened in Geneva in 1974. In 1977 two protocols were adopted. The first deals with the protection of victims of international armed conflicts, and conflicts where people are fighting against colonial domination and alien occupation and against racist regimes in exercise of the right to self-determination; the second deals with armed conflicts not of an international character.

The International Law of Armed Conflict

Unlawful Conduct

The St Petersburg Declaration of 1868 was a milestone in the development of the International Law of Armed Conflict. The general principle embodied in its preamble stated:

> That the only legitimate object which states should endeavour to accomplish during war is to weaken the military forces of the enemy. That for this purpose it is sufficient to disable the greatest possible number of men: That this object would be exceeded by the employment of arms which uselessly aggravate the sufferings of disabled men, or render their deaths inevitable.

Protection is now given to three groups of people: members of military forces who are shipwrecked, wounded or sick; prisoners of war; and civilians. The latter were first specifically covered by the Geneva Convention IV. This *Convention Relative to the Protection of Civilian Persons in Time of War*, under Article 14, places a duty on a belligerent to respect hospitals and safety zones for the wounded, the sick and aged and children, who are all categories of persons afforded special protection. Article 18 provides for the complete protection from attack of civilian hospitals and protection is also afforded to medical services. Section III prohibits the extermination, maltreatment or humiliation of protected persons and prohibits, under Article 33, reprisals against protected persons or their property.

The 1949 Geneva Conventions were amplified in 1977 by two important Protocols. Of particular importance is *Protocol I Relating to the Protection of Victims of International Armed Conflicts*. Article 48 of this Protocol draws a distinction between civilian populations and combatants, and between civilian objects and military objects, and lays down the basic rule that warring parties 'shall direct their operations only against military objectives'. It defines civilians as any persons who do not belong to the armed forces and states that in cases of doubt as to status a person shall be considered to be a civilian. The civilian population comprises all persons who are civilians, and the protection afforded to it is not nullified if there happen to be some non-civilians present among them. Under Article 51 the Protocol affords to civilians 'general protection against the dangers arising from military operations'. In particular, it states:

2 The civilian population as such, as well as individuals, shall not be the object of attack. Acts or threats of violence, the primary purpose of which is to spread terror among the civilian populations are prohibited...

4 Indiscriminate attacks are prohibited. Indiscriminate attacks are:
(a) those which are not directed at a specific military objective;
(b) those which employ a method or means of combat which cannot be directed at a specific military objective; or
(c) those which employ a method or means of combat the effects of which cannot be limited as required by this Protocol:

and consequently, in each case, *are of a nature to strike military objectives and civilians or civilian objects without distinction*. [My emphasis]

5 Among others, the following types of attack are to be considered as indiscriminate:
(a) an attack by bombardment by any method or means which treats as a single military objective a number of clearly separated and distinct military objectives located in a city, town, village or other area containing a similar concentration of civilians or civilian objects; and
(b) an attack which may be expected to cause incidental loss of civilian life, injury to civilians, damage to civilian objects, or

a combination thereof, which would be excessive in relation to the concrete and direct military advantage anticipated.

6 Attacks against the civilian population or civilians by way of reprisal are prohibited.

Under Article 52, general protection from attack or reprisal is accorded to civilian objects and attacks are to be 'limited strictly to military objectives'. The latter are defined as 'objects which by their nature, location, purpose or use make an effective contribution to military action and whose total or partial destruction, capture or neutralization, in the circumstances ruling at the time, offers definite military advantage'. Article 57 lays down rules concerning the planning of an attack and requires that the likely effect of the attack on the civilian population and on individual civilians should be taken into account. It also requires that an attack be cancelled or suspended if it becomes clear that the objective is not a military one or if the attack 'may be expected to cause incidental loss of civilian life, injury to civilians, damage to civilian objects or a combination thereof, which would be excessive in relation to the concrete and direct military advantage anticipated'.

The practical observance of the laws of war is, of course, another matter. The most serious challenge has come from those who have questioned the relevance of the international law of armed conflict to the conduct of modern war, particularly as premeditated destruction of a city or whole society is implicit in theories of strategic bombing.

At the beginning of the Second World War the Governments of Britain and France issued a joint Declaration on 2 September, 1939 affirming their intention to confine aerial bombardment to 'strictly military objectives in the narrowest sense of the word' (Builder and Graubard, 1982, p. 25). The German Government responded on the 17 September by announcing its intention to adhere to the same policy, subject to reciprocity (Builder and Graubard, 1982, p. 25). The aim was to spare the civilian population. However, on 23 September the indiscriminate bombing of Warsaw marked a clear departure from this policy on the part of the Germans, and similar 'violations' occurred on 14 May 1940 with the German bombing of Rotterdam and on the 6 April 1941 with the bombing of Belgrade during 'Operation Punishment' (initiated by the German High Command following the refusal of Yugoslavia to ally herself with Germany).

The accidental bombing of London by German planes on 24 August 1940, as a result of a navigational error, led to a retaliatory British raid on Berlin and to Hitler's decision to blitz London. The basic problem of poor accuracy, coupled with the targeting of industrial concentrations, led to the practice of target-area bombing. It soon became clear that those directing the bombing campaigns were beginning to justify the targeting of civilian populations as an end in itself. The British Directive on Air Warfare of 29 October 1942 in theory limited bombardment to military objectives and forbade the intentional bombing of civilian populations. However, it removed limitations in respect of *all* German, Italian or Japanese territories on the grounds that 'consequent upon the enemy's adoption of a campaign of unrestricted air warfare the Cabinet have authorised a bombing policy which includes the attack on enemy morale' (Builder and Graubard, 1982, p. 22). Similarly, in the Casablanca Directive of 21 January 1943 the combined Chiefs of Staff defined the objective of the bomber offensive against Germany as being the 'progressive destruction and dislocation of the German military, industrial and economic system, and the undermining of the morale of the German people to a point where their capacity for armed resistance is fatally weakened' (cited in Brooks, 1983, p. 35).

In practice the list of legitimate targets now includes centres of communications, large industrial and administrative establishments of any kind, and any area likely to become important for the conduct of war. Modern targeting policies have blurred the distinctions upon which The Hague and Geneva Conventions are based, while the technology of war has made the distinction between military and civilian personnel and objects difficult if not, in the case of nuclear weapons, impossible, to maintain. Indeed the fact that no distinction is possible is a major reason for maintaining that nuclear weapons should be banned. As Morris Greenspan argues (1959, p. 371), 'it is essential that lawful weapons should be sufficiently selective in their effects to allow for such differentiation'. However, it might be argued that while observance of the laws of war would prohibit a nuclear first-strike that encompassed civilian targets or, by its nature, affected civilians, a retaliatory strike might be allowable. This leads to a consideration of the international law on reprisals.

Reprisals in war are the commission of acts which, although unlawful in themselves, may under certain specific circumstances and in specific instances, be justified because the adversary has himself behaved unlawfully. National war codes generally recognize the possibility of resorting to reprisals. Attempts to regulate them through International Law, as at the Brussels Conference of 1874, failed, and on this subject customary law is vague and indefinite. In general it is assumed that reprisals can only be resorted to as a means of persuading the adversary to observe the laws of war in the future, and not to obtain revenge or to punish. It is accepted that the means employed need not be the same as those used in the original offence. It used to be accepted that non-combatants had no claim to exemption from reprisals. Indeed the burning of towns and villages has been a common form of reprisal. However, under the Geneva Convention IV Article 33 the civilian population and civilians were specifically excluded. This prohibition was made quite explicit in Article 51 of the Geneva Protocol I which says 'Attacks against the civilian population or civilians by way of reprisals are prohibited.'

Unlawful Weapons

The St Petersburg Declaration of 1868 was of particular importance because it was the first international instrument in the modern period that prohibited a weapon. Under the Declaration the contracting parties agreed to renounce small exploding and incendiary bullets. Subsequent international agreements have outlawed other weapons. These include the Hague Declaration of 1899 which outlawed dum-dum bullets, and a further convention of 1981 which opened for signature on 10 April 1981 and which seeks to enforce prohibitions or restrictions of use on landmines, booby-traps, incendiary weapons and weapons producing fragments not detectable by X-ray. Early efforts were reinforced by the Hague Convention IV of 1907 which, in the Annex to the Convention, Section II, Chapter I states in Article 22 that 'The right of belligerents to adopt means of injuring the enemy is not unlimited,' and in Article 23 especially forbids the use of poison or poisoned weapons and the use of 'arms, projectiles or material calculated to cause unnecessary suffering'.

Particularly important in the development of international laws banning the use of certain weapons is the *Protocol for the Prohibition of the Use in War of Asphyxiating, Poisonous or Other Gases and of Bacteriological Methods of Warfare* (1925). This Protocol, normally referred to as the 'Gas Protocol', has become a universally binding instrument, at least as far as first use of any prohibited weapon is concerned. The continued adherence of nations to this Protocol is shown by the fact that the USA ratified it as recently as 1975. The existence of the Protocol has not prevented countries from maintaining stocks of chemical weapons, including poison gases, although Britain, having destroyed its own stocks of chemical weapons, proposed to the UN Committee on Disarmament in February 1982 that measures be taken to ensure that stocks of chemical weapons would be destroyed and that no future production would be carried out either overtly or in secret.

Like the Hague Convention IV of 1907, Protocol I to the Geneva Conventions (1977) also states, in Article 35, that the right of belligerents to 'choose methods or means of warfare is not unlimited'. It prohibits the use of weapons likely to cause 'superfluous injury and unnecessary suffering', and also prohibits 'methods or means of warfare which are intended, or may be expected, to cause widespread long term and severe damage to the natural environment'. This latter prohibition is further developed in Article 55 which states:

1 Care shall be taken in warfare to protect the natural environment against widespread, long term and severe damage. This protection includes prohibition of the use of methods or means of warfare which are intended or may be expected to cause such damage to the natural environment and thereby to prejudice the health or survival of the population.
2 Attacks against the natural environment by way of reprisals are prohibited.

The Protocol also, in Article 36, enjoins parties to it to determine the legality of the use of any new weapon which may be invented. To this extent it is timeless.

The effects of nuclear weapons, particularly in respect of initial radiation and fall-out, place them in a similar category to weapons prohibited under the Hague Conventions of 1899 and 1907, and under the Gas Protocol of 1925. The importance of the latter is that it was treated in the inter-war years as declaratory of existing international law. This is tantamount to total prohibition of the use of such weapons except under the limited conditions laid down for the use of prohibited weapons in reprisals against enemy military forces. As yet there are no specific international agreements outlawing or limiting the use of nuclear weapons. Indeed, Britain, while it signed the Geneva Protocol I, purported to make a reservation to its signature that it does not regard agreement as affecting the legality of nuclear weapons.

Notwithstanding such equivocations, there have been a number of condemnations of nuclear weapons. The UN General Assembly Resolution 1653 (XVI) of November 1961, which was adopted by a vote of 55 to 20 with 26 abstentions, declared that:

> 1 (a) The use of nuclear and thermo-nuclear weapons is contrary to the spirit, letter and aims of the United Nations, and as such, a direct violation of the Charter of the United Nations.
> (b) The use of nuclear and thermo-nuclear weapons would exceed even the scope of war and cause indiscriminate suffering and destruction to mankind and civilization and as such, is contrary to the rules of international law and the laws of humanity.
> (c) The use of nuclear and thermo-nuclear weapons is a war directed not against an enemy or enemies alone but also against mankind in general, since the peoples of the world not involved in such a war will be subjected to all the evils generated by the use of such weapons.
> (d) Any State using nuclear and thermo-nuclear weapons is to be considered as violating the Charter of the United Nations, as acting contrary to the laws of humanity and as committing a crime against mankind and civilization.

Similar resolutions in 1978 and 1980 have reaffirmed these humanitarian standards. Exactly what weight such resolutions have is disputed. The reason most frequently advanced for not accepting UN General Assembly Resolutions as capable of establishing binding principles and rules, beyond the very limited rule-making powers provided in the Charter of the UN, is that the states voting for the resolutions do not make institutional arrangements confirming their commitment. On the other hand, Professor Eric Suy, Legal Counsel of the UN, has claimed that 'the General Assembly resolution is becoming a useful modern tool for standard-setting and rule-creation in an expanded international society that requires more rapid formulations of standards governing the conduct of its members' (Suy, 1981, p. 11). Various writers such as Professor Richard Falk and Gaetano Arangio-Ruiz have suggested that the Assembly should be regarded as creating 'weak' legislative norms (Falk, 1970, p. 180) or 'a law determining interpreting and developing function in a "non-technical" sense' (Arangio-Ruiz, 1979, p. 77). While there is no consensus, there are many people who would agree with Rosalyn Higgins (1963, p. 5) that while 'Resolutions of the Assembly are not *per se* binding... the body of resolutions as a whole, taken as indications of a general customary law, undoubtedly provide a rich source of evidence.'

Some commentators argue that since there are no international instruments prohibiting the possession, testing (under certain conditions), or use of nuclear weapons, they are not illegal. The position of the USA, for example, has been that nuclear weapons are subject to the same principles which apply to other weapons for their lawful use in armed conflict. Others argue that the weight of the international law of armed conflict, either directly or indirectly by analogy makes the use of nuclear weapons illegal, and that this is supported by specific treaties.

The Treaty of Tlatelolco (1967) established a regional nuclear-free zone in Latin America. The Non-Proliferation Treaty, while it can be held to implicitly recognize the existence of nuclear-weapon states, could also be regarded as an instrument for establishing, by customary law, a renunciation of nuclear weapons. Thus Article 1 of the Treaty pledges nuclear weapon states not to transfer to non-nuclear weapon states or 'to any recipient whatsoever' nuclear weapons, or control over them, either directly or indirectly. Article 6 pledges parties to pursue and effect complete disarmament. Under Article 2, non-nuclear-weapon states agree not to receive 'from any transfer whatsoever' nuclear weapons or control over them.

Implications for Nuclear Weapons

It seems clear, despite the protestations of nuclear-weapon states to the contrary, that these weapons are illegal for the following reasons:

1 They infringe the Hague Conventions (1899 and 1907) and the Gas Protocol (1925) since they create radioactivity, which is

a poison. It is argued by some proponents of nuclear weapons that, in the absence of a specific prohibition against the use of nuclear weapons, they must be regarded as legal. However the absence of a specific injunction does not necessarily imply that they are not covered by International Law. Article 36 of the Geneva Protocol I (1977) requires states to determine the legality of the use of any new weapons which may be invented.

2 They cause or may be expected to cause widespread, long-term and severe damage to the environment, contrary to Article 35 of the Geneva Protocol I.

3 Contrary to the 1907 Hague Convention, they violate the territory of neutral countries since, at the very least, such countries suffer the direct effects of fall-out.

4 Except in the very limited cases of their use against naval ships at sea or in 'surgical strikes' against geographically remote and isolated military bases and targets, they do not and cannot distinguish between combatants and non-combatants, contrary to the 1949 Geneva Conventions and particularly the 1977 Geneva Protocol I.

5 They may, under certain conditions (a retaliatory strike against cities) be held to contravene the 1949 Geneva Conventions and the Geneva Protocol I, which forbids attacks on civilians as a reprisal.

6 Except in the very limited cases mentioned in 4 above, they contravene the legal provisions designed to protect property and buildings including civilian hospitals, institutes of art, medicine, science or religion, and historic or charitable institutions.

7 They may aggravate the suffering of disabled men, contrary to the Declaration of St Petersburg and the 1949 Geneva Conventions.

8 They may in certain circumstances be construed as an act of genocide, contrary to the Convention on Genocide.

So far as the production and possession of nuclear weapons is concerned, there are those who argue that since the use of nuclear weapons is illegal (from the above), acts preparatory to their use are similarly illegal. Article 36 of the Geneva Protocol I charges those responsible for the development and procurement of weapons with a duty to ensure that the weapons acquired are suitable for use under international law. Since the nuclear weapons states have judged that they are suitable, it is unlikely that they would freely interpret this Article as applying to their weapons' programme. Nevertheless, the weight of evidence must support the opposite contention – that since the use of such weapons is illegal their development and procurement is also contrary to the spirit of this Article. By the same token, the Nuremberg Tribunal found civilians who had supplied the Nazis with Zyklon B guilty of war crimes. It is notable that directives of the US Department of Defense require that all persons responsible for the acquisition and procurement of weapons ensure that their actions are consistent with the international law of armed conflict.

At another level, it can be argued that renunciation of the use of poisonous and biological weapons does not prevent states from acquiring such weapons as an insurance policy and making use of them as a deterrent which they can legally threaten to use in retaliation. International lawyers argue that it is not at all clear that the possession of nuclear weapons is forbidden. However, the Non-Proliferation Treaty clearly enjoins non-nuclear-weapon states not to acquire such weapons, and all parties (but especially the nuclear-weapon states) to negotiate arms-control agreements leading to complete nuclear disarmament. In addition to the obligation (which derives from Article 36 of the Geneva Protocol I) not to procure weapons unsuitable for use, it is explicit in the Non-Proliferation Treaty and implicit in international law that the theory of arms control should as Builder and Graubard (1982, p. 49) put it, derive 'from concepts which impute lawful behaviour by both parties in compliance with the international law, not only with arms control treaties as the law, but with the laws governing the use of weapons of war'.

Finally, there is the question of the targeting of nuclear weapons. In a real sense, the theory of deterrence involves holding the entire population of another country hostage. What is new is that nuclear weapons have enabled this to be done without any obvious restraint being placed on the hostages. The fact that we are all hostages can be forgotten. Article 51 of the 1977 Geneva Protocol I prohibits 'Acts or *threats of violence* [my italics] the primary purpose of which is to spread terror among the civilian population'. That those living under the threat of nuclear attack can forget or ignore it makes it no less real. The implication of this Article is that any targeting policy which poses a threat to the civilian population or to civilians is itself unlawful, and this will be the case whether or not the attack is a first-strike or a retaliatory second-strike.

Thus at the level of development and procurement, of possession, targeting and use, the weight of the international law of armed conflict is against the legality of nuclear weapons, and it becomes a responsibility of a nation's citizens to ensure that its government honours its responsibilities in international law.

The Status of International Law

Well into the Middle Ages, the rules covering conduct between states were thought to be so obvious and elementary that they were identified with the rules of nature (*jus naturale*). As a result of the spread of religious concepts, the law of nature came to be equated with Divine Law in the writings of both Christian and Moslem theologians as well as naturalists (as they came to be called). Law, including international law, was viewed as part of an immutable body of precepts waiting to be discovered.

As differing interpretations of the law multiplied, so the idea that the law was immutable came to be challenged. Positivists such as Cornelius van Brykershock, Johann Jacob Moser and George Friedrich von Martens came to believe that the sources of international law are international obligations explicitly undertaken by states. In the absence of a corpus of international agreements to guide the resolution of conflicts, it was, however, difficult to adhere to the positivist principles and, as a result, a compromise school emerged that tried to bridge the gap between naturalists and positivists. The Grotians (named after Hugo Grotius) drew a distinction between *necessary law* (the law of nature and of God) and *voluntary law* (consisting of treaties and customs made by man to meet particular circumstances). These concepts still form the basis of international law. Article 38 (1) of the 1945 Statute of the International Court of Justice designates the terms of reference of international law to be:

a) international conventions, whether general or particular, establishing rules expressly recognised by the contesting states;

b) international custom, as evidence of a general practice accepted as law;
c) subject to the provisions of Article 59, judicial decisions and the teachings of the most highly qualified publicists of the various nations, as subsidiary means for the determination of rules of law.

Although no hierarchy of precedence is attributed to (a), (b) and (c) of Article 38 (1), the West European tradition of international law cites custom as the most important source of international law. Professor Johnson (1981, p. 23), for example, stated that 'the first and most important source of this "international law" is the custom of states'. The status and content of customary law has been heavily criticized, particularly by non-Western European states. Criticism stems from its close affinity with Western European values and legal systems. Also in dispute is the length of time required to elapse before a practice becomes customary. The original fiction was that the rules had been observed 'for time immemorial', but contemporary writers speak of the possibility of 'instant custom' arising from UN resolutions. In the *South West Africa* case at the International Court of Justice, Judge Tanaka in a dissenting Opinion, suggested that 'the establishment of such a custom would require no more than one generation or even far less than that' (ICJ, 1966, p. 291).

It is also not clear what the requirements are regarding proof of the belief that a legal obligation exists. In the *North Sea Continental Shelf* case the International Court said:

> Not only must the acts concerned amount to a settled practice, but they must also be such, or be carried out in such a way, as to be evidence of a belief that this practice is rendered obligatory by the existence of a rule of law requiring it. The need for such a belief, i.e. the existence of a subjective element, is implicit in the very notion of the *opinio juris sive necessitatis*. The States concerned must therefore feel that they are conforming to what amounts to a legal obligation (ICJ, 1969, p. 144, para. 77).

This may not necessarily require a state to explicitly enter into an agreement. It has been suggested that, 'providing a State does not expressly object to a "norm creating" treaty in force among other States, and does not act in contravention of that treaty, it may be bound by the rules established in the treaty in the form of a rule of international customary law' (Dahlitz, 1983, p. 97). As Rosas (1981, p. 5) put it: 'It seems pertinent to speak of *presumed* consent, which means that a State can be bound by a rule the creation of which it has not participated in as long as it has not clearly voiced its opposition.' There is, of course, a great difference in principle between a state *acting as if* bound by a treaty, and *giving its explicit consent* to be bound.

Analysing the ICJ's Opinion in the *North Sea Continental Shelf* case, Dahlitz (1983, pp. 105–6) identifies the criteria invoked by the Court as follows:

1 The rule must be sufficiently significant to be of a *fundamentally* norm-creating character (ICJ 1969, p. 41, para. 72).
2 The rule must be of a *potentially* norm-creating character with respect to precision as to its meaning and scope (p. 42).
3 State practice constituting the subject matter of the rule should be extensive and virtually uniform (p. 43, para. 74).
4 The passage of a considerable period of time is not necessary, provided there is widespread and representative participation in the observance of the rule, particularly among States having a direct interest in the issue (p. 42, para. 73).
5 The States that adopt the practice constituting the rule must do so because they feel legally compelled to comply (p. 44, para. 77).
6 The States that observe the rule because they feel legally compelled to do so, must believe themselves to be bound by a custom as representing law, not merely by treaty obligation (pp. 44–5, para. 77).
7 The treaty embodying the rule in question must have been, at its inception, declaratory of a then existing rule of customary international law (p. 45, para. 81).

The International Law of Armed Conflict

The basic principles guiding the international law of armed conflict are inspired by humanitarian considerations which have their roots in Christian and European concepts of chivalry, professional honour and common humanity to those in less fortunate circumstances. Such principles may not be self-evident in all cultures.

Tribal loyalty may enjoin the destruction of enemies; the concept of a holy war may make the destruction of infidels appear a duty.

In part, basic humanitarian principles may accord with concepts of military necessity. A US Air Force pamphlet AFP 110-31 (1976) suggests that use of excessive force and infliction of wanton destruction and unnecessary suffering on the enemy not only wastes military resources but violates developed legal principles (p. 5). It goes on to point out that experience derived from the Second World War suggested that 'coercive warfare' such as the bombing of cities was 'militarily ineffective' (p. 5).

Another question concerns the justification of particular military actions on the grounds of necessity. The problem is to establish which acts are forbidden by the laws of war. These are distinguished from those forbidden by the usages of war. Acts forbidden by law can never be committed even in cases of military necessity. Acts which are forbidden by usage can be committed if it is militarily necessary to do so. However, where military necessity cannot be invoked, the 'Martens Clause' of the Hague Convention IV (1907) holds. This states that 'Until a more complete code of the laws of wars has been issued ... the inhabitants and the belligerents remain under the protection and the rule of the principles of the law of nations, as they result from the usages established among civilized peoples, from the laws of humanity, and the dictates of the public conscience.'

International conventions such as the Hague and Geneva Conventions are of great importance in assessing the legality of particular methods of waging war. If sufficient states accede to the treaty, then the laws embodied in it may become what is known as customary law which is then deemed to be binding on all states, and is part of what is referred to as the 'common law' of nations. In the Nuremberg judgement in the case of Major War Criminals, the International Military Tribunal observed: 'The law of war is to be found not only in treaties, but in the customs and practices of states which gradually obtained universal recognition, and from the general principles of justice applied by jurists and practised by military courts. This law is not static, but by continued adaptation follows the needs of a changing world.' Customary law may thus supplement and modify legislation which is itself often only a codification of principles already generally accepted. The writings of lawyers and judicial decisions may also influence the development of customary law and the interpretation of the principles of law embodied in treaties.

In general, international law serves to regulate the conduct of states, but in the case of the international law of armed conflict it also regulates the actions of individual combatants. Any person, whether a civilian or a member of an armed force, who commits an act that constitutes a crime under the international law of armed conflict is liable to punishment. What constitutes such a crime is defined by Article 85 of the 1977 Protocol I to the 1949 Geneva Conventions:

> 3 In addition to the grave breaches defined in Article II [dealing with medical abuses], the following acts shall be regarded as grave breaches of this Protocol, when committed wilfully, in violation of the relevant provisions of this Protocol, and causing death or serious injury to body or health:
> (a) making the civilian population or individual civilians the object of attack;
> (b) launching an indiscriminate attack affecting the civilian population or civilian objects in the knowledge that such an attack will cause excessive loss of life, injury to civilians or damage to civilian objects...

Such grave breaches are regarded as war crimes.

Of particular importance in the interpretation of breaches of the international law of armed conflict are the *Nuremberg Principles* which were formulated by the International Law Commission in 1950. Principle VI provides:

> The crimes hereinafter set out are punishable as crimes under international law:
> (a) crimes against peace:
> (i) planning, preparation, initiation or waging of a war of aggression or a war in violation of international treaties, agreements or assurances;
> (ii) participation in a common plan or conspiracy for the accomplishment of one of the acts mentioned under (i).
> (b) war crimes;
> violations of the laws or customs of war which include, but are not limited to ... wanton destruction of cities, towns or villages, or devastation not justified by military necessity.

The *British Manual of Military Law* makes it quite clear that 'members of the armed forces are bound to obey lawful orders only and they cannot escape liability if, in obedience to command, they commit acts which both violate the unchallenged rules of warfare and outrage the general sentiment of humanity' (Ch. 14, para. 443). Both British and American servicemen are required to observe the international law of armed conflict.

Returning to the seven criteria invoked by the International Court of Justice in the *North Sea Continental Shelf* case (see p. 194), it seems clear that:

1 The international law of armed conflict as laid down in, for example, the Hague Conventions (1899 and 1907), the Geneva Conventions (1949), the Geneva Protocol (1925) and the Geneva Protocols (1977), does constitute a body of rules that are of sufficient significance to be fundamentally norm-creating.
2 The international law of armed conflict as spelt out in these Conventions and Protocols is sufficiently precise to be judged to be norm-creating.
3 State practice in respect of adoption of the international law of armed conflict is extensive and virtually uniform. Thus UN General Assembly 2677 (XXV) of 9 December 1970, which called on all parties to any armed conflict to observe the rules laid down by the Hague Conventions of 1899 and 1907, the Geneva Protocol of 1925, the Geneva Conventions of 1949, and other humanitarian rules applicable to armed conflicts, was adopted by 111 votes to nil, with 4 abstentions.
4 While not necessary to establish their legitimacy as customary law, many of the rules embodied in the international law of armed conflict have been accepted for a considerable period of time.
5 States have felt legally compelled to comply, at least in theory, with the rules established by the international law of armed conflict, and have condemned those states that have not complied with them.
6 States have commonly regarded the international law of armed conflict as embodying customary law and not as an obligation arising from treaty.
7 The treaties embodying the rules were declaratory of existing customary law.

While the application of these international laws to nuclear weapons has not been accepted by nuclear-weapon states, there is little doubt that the use of nuclear weapons contravenes the existing provisions of the international law of armed conflict. The General Assembly of the UN has a role in creating norms, both through its Special Sessions (of which there have been two on Disarmament) and through its resolutions. In its Resolution 1653 (XVI) of November 1961, and in subsequent resolutions in 1978 and 1980, it has been established that there is a considerable body of international opinion against the use of nuclear weapons, and a belief that their use would contravene the rules of international law and the laws of humanity. There is also a view, to which reference has already been made, that the resolutions of the Assembly 'taken as indications of a general customary law', provide a rich source of evidence that may inform the development of international law. Given this, there is an urgent need to reinforce the norm-creating and legislative functions of the UN, and to rectify the deficiencies of the present system of international law, as a support to efforts for nuclear arms control and disarmament. The particular case that nuclear weapons if used would contravene existing international law is well founded.

References and Further Reading

Arangio-Ruiz, G. (1979) *The United Nations Declaration on Friendly Relations and the System of the Sources of International Law*, Maryland: Sijthoff and Nordhoff.

Brooks, S. (1983) *Bomber. Strategic Air Power in Twentieth Century Conflict*, London: Imperial War Museum.

Builder, C.H. and Graubard, M.H. (1982) *The International Law of Armed Conflict: Implications for the Concept of Assured Destruction*, Santa Monica: Rand Corporation Document, R-2804-FF.

Dahlitz, J. (1983) *Nuclear Arms Control*, London: George Allen & Unwin.

Falk, R. (1970) *The Status of Law in International Society*, Princeton: Princeton University Press.

Glahn, G. von (1970) 2 edn, *Law among Nations: An Introduction to Public International Law*, New York: Macmillan.

Greenspan, M. (1959) *The Modern Law of Land Warfare*, Los Angeles: University of California Press.

Higgins, R. (1963) *The Development of International Law through the Political Organs of the United Nations*, London: Oxford University Press.

International Court of Justice, Cases:
North Sea Continental Shelf Cases (Federal Republic of Germany/Denmark; Federal Republic of Germany/Netherlands), I.C.J. Rep. 1969.
South West Africa Case (Ethiopia v. South Africa; Liberia v. South Africa), I.C.J. Rep. 1966.
International laws relating to war and armed conflicts;
The Declaration of St Petersburg, 1868.
The Hague Conventions, 1899:
Declaration IV (2) Concerning Asphyxiating Gases.
Declaration IV (3) Concerning Expanding Bullets.
The Hague Conventions, 1907:
Convention (IV) Respecting the Law and Customs of War on Land.
Protocol for the prohibition of the use in war of asphyxiating, poisonous or other gases, and of bacteriological methods of warfare, 1925 (Geneva Protocol or Gas Protocol).
Convention on the Prevention and Punishment of the Crime of Genocide, 1948 (Genocide Convention).
Conventions for the Protection of War Victims, 1949 (Geneva Conventions):
Convention (I) Providing for the amelioration of the condition of the wounded and sick in the field.
Convention (II) Providing for the amelioration of the condition of the wounded, sick and shipwrecked members of armed forces at sea.
Convention (III) Relative to the Treatment of Prisoners of War.
Convention (IV) Relative to the Protection of Civilian Persons in Time of War.
Protocol Additional to the Geneva Conventions of 12 August 1949 and relating to the protection of victims of international armed conflicts, 1977 (Geneva Protocol I).
Protocol Additional to the Geneva Conventions of 12 August 1949 and relating to the protection of victims of non-international conflicts, 1977 (Geneva Protocol II).
Convention on the prohibition of military or any other hostile use of environmental modification techniques, 1977 (ENMOD Convention).
Convention on the prohibition or restriction on the use of certain conventional weapons which may be deemed to be excessively injurious or to have indiscriminate effects, 1981.
International Treaties:
Treaty for the Prohibition of Nuclear Weapons in Latin America, 1967 (Treaty of Tlatelolco).

Treaty on the Non-Proliferation of Nuclear Weapons, 1968 (Non-Proliferation Treaty).

International Military Tribunal, Nuremberg. Quotation from Shawcross in Volume 19, p. 427.

Johnson, B. (1981) 'Changes in the Norms Guiding the International Legal System – History and Contemporary Trends', *The Spirit of Uppsala* (JUS 81). Manuscript No. 44. Papers of the Joint UNITAR–Uppsala University Seminar on International Law, 1981.

Kent, B. (1983) 'Notes, nuclear, servicemen for the use of', *The Guardian*, 14 May.

Lawyers versus the Bomb (n.d. c. 1982) *The Illegality of Nuclear Weapons*, London: Lawyers versus the Bomb.

Miller, R.I. (1975) *The Law of War*, Lexington, Mass: D.C. Heath.

Oppenheim, L. (1961) 7 edn, *International Law: a Treatise*, edited by H. Lauterpacht, New York: David McKay Co. Ltd., vol. II. *Disputes, War and Neutrality*.

Rosas, A. (1981) 'Customary Law: from "Universal" in a European System to "Regional" in a World System'. *The Spirit of Uppsala* (JUS 81). Manuscript No. 56. Paper of the Joint UNITAR–Uppsala University Seminar on International Law, 1981.

Schwarzenberger, G. (1958) *The Legality of Nuclear Weapons*, London: Institute of World Affairs/Stevens & Sons Ltd.

SIPRI (Stockholm International Peace Research Institute) (1978) *Arms Control: A Survey and Appraisal of Multilateral Agreements*, London: Taylor & Francis.

Suy, E. (1981) 'A New International Law for a New World Order'. *The Spirit of Uppsala* (JUS 81). Manuscript No. 43. Papers of the Joint UNITAR–Uppsala University Seminar on International Law, 1981.

United States Air Force (1976) *International Law – The Conduct of Armed Conflict and Air Operations*. Document AF 110-31, 19 November 1976.

9
The Ethics of Nuclear War and Deterrence

What is Ethics?

The use of nuclear weapons, even the intention or threat to use them, raises ethical or moral issues which have been extensively debated by moral philosophers and theologians. Moral philosophers broadly divide into two groups. One believes that ethics is a system of rules that ought to govern our actions. Judgements about what one should do in particular instances are always determined in the light of these rules; an act is right not because it has good or bad consequences, but because the action itself has some feature that makes it right. For example, it is just, it keeps a promise, or it is commanded by God. The other group of moral philosophers says that it is not the action or the rule that is good or bad but the consequences deriving from it. Moral philosophers who believe this are called *consequentialists* (or *teleologists*), and are contrasted with rule-based moral philosophers, who are called *deontologists*. The best known consequentialist theory is that of utilitarianism which in its classical form states that an action is right if it produces as much or more happiness or good for everyone affected by it than does any alternative action, and wrong if it does not. It follows that there are no universally applicable rules. Circumstances can alter the judgements we make.

Most moral philosophers are agreed that it is not enough to justify one's actions in terms of self-interest alone, nor in terms of any partial or sectional group, although there are exceptions to this. Generally however, moral philosophers agree that ethics takes a universal point of view and that one's own interests cannot carry more weight than any others. The basic distinction between deontologists and consequentialists is, however, important since it influences the way in which individual moral philosophers approach the ethics of nuclear war and deterrence.

Christian Arguments

Christian arguments are predicated on a number of assumptions, including a belief in the existence of God, and a belief that God has revealed, through a framework of rules (commandments), and by example (particularly through the life of Jesus, but also through the Saints), the principles by which mankind should live. One argument, which figures prominently in the recent Church of England working party report on *The Church and the Bomb* (CEBSR, 1982) is that mankind has a God-given responsibility towards God's creation. Men and women exercise a trusteeship over creation which enables them to participate in God's plan. In exercising this trusteeship, they are free to commit good or evil (Genesis, 2–3) (p. 106). To help them God revealed His Law and sent Jesus to live among them.

One of the issues which Christians have to face is the moral dilemma posed by war and the weapons of war. Christians are expected to place a high value on peace. In the Old Testament peace is seen as part of the expectation of the messianic age to come. In the New Testament, it is a blessing to be sought in the present (1 Peter, 3:11), and flows from the peaceful temper that is the special mark of a disciple of Christ. Christians have a spiritual duty to be peacemakers which goes beyond mere co-existence to the promotion of social and economic justice within and between nations, the opening up of a dialogue between those with different beliefs, and the support of institutions designed to promote and secure peace (CEBSR, 1982, pp. 111–3). Nuclear weapons are a direct denial of the Christian concept of peace since their existence encourages fear and undermines mutual trust and respect between peoples (CEBSR, 1982, p. 148).

Some Christian theologians have argued that the destruction of the world might not be the worst evil imaginable. The Roman Catholic Archbishop Oscar Lipscomb of Mobile, Alabama, expressed one version of this view:

> We seem to assign the human species itself a right to eternity. This is certainly not the 'Biblical vision of the world at the heart of our religious heritage'. The worst evil that can befall us is not the loss of our life, or even of all human life. It is sin and the consequent loss of that life in the Father through Christ by means of the Spirit that we rightly call 'life everlasting'. Should this world and our species remain in such a way that such life in the Father is not possible to the generations that would follow, then we have threatened not just the sovereignty of God over the world, but the victory of Christ over sin and death (Lipscomb, 1982).

This view is not widely held and is directly challenged by the authors of *The Church and the Bomb*, who argue that while it is true that this world is not, for Christians, the whole story, Christians 'are not entitled to be selective in quoting Jesus to support [their] arguments' (CEBSR, 1982, p. 146). With equal force the Gospel places on Christians 'the duty of peacemaking; compassion for the unfortunate and the deprived; the proclamation of Jesus as victor over sin, disease and death; the urgency of converting others to Christ's ways, and not letting them end their lives here in ignorance of God's love' (p. 146).

Pacifism

Absolute Pacifism

Some individuals argue that any resort to force or violence is evil, and that in no circumstances should one kill another human being. This belief is sometimes based on a secular view of the sanctity of human life, and sometimes on religious belief. Christian pacifists, for example, point to the Commandment against killing and the example of Jesus who refused to defend himself against his enemies. They point out that Jesus taught his followers to love their enemies and to return good for evil by turning the other cheek when assaulted. Pacifism, they argue, must be as integral a part of the Christian's life as it was of Jesus's. Some pacifists argue that all professing Christians should be pacifists since pacifism is central to Christian faith. Others see it as a vocation which some Christians are called to exercise. Holding pacifist views does not prevent the individual from employing the techniques of non-violent resistance and non-cooperation. Some pacifists go further, and distinguish between the use of *force*, to restrain, and *violence*. All absolute pacifists regard killing, even in self-defence, and whatever the consequences, as morally wrong.

Some critics of pacifism argue that the pacifist does not cater for the situation in which a violent act now may prevent greater violence in the future. Here the choice is not between a good action and an evil one, but between a lesser and a greater evil. Others ask what pacifists would do if faced by a choice between killing and letting die, as, for example, in the case where an assailant is threatening to kill an innocent passer-by. Assuming restraint is impossible should one kill the assailant if one has the means to, rather than let the innocent person be killed? Such critics ask what right pacifists have to place a higher value on their own integrity as non-violent people than on the saving of life. The problem with personal integrity, as Harris points out (1980, p. 116), is that 'it does not help us with moral dilemmas, for the dilemmas... are about what price to put on our integrity'. Pacifists reply, however, that their refusal to intervene violently is itself a positive act. By remonstration, by the passive acceptance of the violence done, by resisting through non-violent means, one can hope to shame the aggressor into a renunciation of violence.

Consequential Pacifism

Another form of pacifism is based on utilitarian principles rather than on any rigid rule against the use of violence. Such consequential pacifists argue that establishing and maintaining a tradition in which war is repudiated has far more beneficial consequences than the most desirable outcome of any war, however just one's cause. Every resort to war helps to legitimatize war as an activity, but every time one refuses to take part in a war, one contributes to the establishment of a tradition for using alternative means for the settlement of disputes. Any recourse to war is therefore more evil than the evil which might follow from not going to war. If this is true of conventional war, it is even more true of nuclear war.

Faced by the possibility of nuclear war and annihilation, Monsignor Bruce Kent has argued that Christians cannot place national survival (as an independent nation) above everything else. They 'have to take seriously the possibility that the way of the Cross is actually a practical political option and that faith as well as prudence may suggest that there are times when it is better to suffer injustice than to defend ourselves. Loving our enemies may mean surrendering our lives or our liberty rather than taking theirs' (Kent, 1982, p. 56).

'Better Red than Dead'

Pacifists are sometimes accused, usually in a derogatory way, of believing that it is 'better to be red than dead'. The US National Security Advisor Richard V. Allen (1981) was reported as saying that 'better red than dead' is a 'contemptible slogan'. Lord Chalfont, writing in *The Times*, argued that 'there can be nothing – *nothing* – worse than a life in which, by the exercise of relentless tyranny, the precious gifts of liberty and dignity are denied' (cited in CEBSR, 1982, pp. 146–7).

In fact, of course, a large number of people do live under communist regimes and most of them clearly find life worth living. This is not to make any judgement about whether life under a communist regime is better or worse than life under a non-communist one; it is merely to say that most people prefer life under a communist regime to death. Moreover, it does not deny anyone whose country is ruled by communists, or overrun by them, the right to resist. Individuals who decide that it would be better to be dead than red can only speak for themselves. As McMahan points out (1981, p. 127) 'those who in the name of freedom, would actually prefer to see Britain destroyed by nuclear weapons than to see it ruled by the Russians are guilty of inconsistency. Supposedly in the defence of freedom, they would deny us the most fundamental freedom: the freedom to determine our fate, to decide whether to live or die'.

Just-War Theory

Just-war theory attempts to lay down criteria by which a war can be judged to be fought in a just cause and manner. As well as forming a basis for the development of the international law of armed conflict (see pp. 179–80), it is also used by Protestant and particularly Roman Catholic theologians to determine the moral basis for going to, and conducting, war. The development of nuclear weapons has posed significant problems for the theory which are still being debated.

The Principle of a Reasonable Probability of Success

Just-war theory requires that those embarking on a war should have a reasonable hope of success. Success implies that the justifiable aims of the war can be attained: that is, that a wrong can be righted, and that peace, justice and order can be restored after the war. The effects of nuclear war are now recognized to be such that no country engaging in a nuclear war can have a reasonable hope of success in any meaningful sense of the word. Pope John Paul II, speaking at Hiroshima, argued that nuclear weapons threaten the whole planet, and 'compel everyone to face *a basic moral consideration: from now on, it is only through a conscious choice, and then deliberate policy that humanity can survive*' (John Paul II, 1981, p. 2, italics in the original). Both the Church of England's report (CEBSR, 1982, p. 96) and the US Catholic Bishop's Pastoral Letter (US Catholic Conference, 1983, p. 43) have condemned initiation of nuclear war.

The Principle of Proportionality

The principle of proportionality states that a war must not result in disproportionate harm for any of the populations engaged in it. This is not merely a question of weighing up the good achieved and the harm caused, it also requires that the action be an appropriate means of achieving a lawful end. The principle affects both the decision whether or not to go to war and the means used. During the Second World War the principle was invoked by Pope Pius XII (1944) who argued that 'the theory of war as an apt and proportionate means of solving international conflicts is now out of date', while Pope John XXIII (1967, para. 127) argued that 'in this age which boasts of its atomic power, it no longer makes sense to maintain that war is a fit instrument with which to repair the violation of justice.' However, while the right to redress a wrong was being challenged, the Roman Catholic Church does not challenge the right of governments 'to legitimate defence once every means of peaceful settlement has been exhausted' (US Catholic Conference, 1983, p. 21).

Protestant theologians have been more circumspect in their condemnation of the right of countries to go to war in a just cause. Angus Dun and Reinhold Niebuhr (1955) argued that 'The notion that the excessive violence of atomic warfare has ended the possibility of a just war does not stand up.' Christians, they said, had a duty to participate in war 'waged to vindicate what they believe to be an essential Christian principle: to defend the victims of wanton aggression, or to secure freedom for the oppressed'. However, under the conditions of 'modern warfare', 'Defense against aggression seems a safer and saner justification than securing the freedom of those already oppressed. Because of the consequence that war today may entail, only the most imperative demands of justice should be honoured.' For these reasons, 'the occasions to which the concept of the just war can be rightly applied have become highly restricted ... The concept of a just war does not provide moral justification for initiating a war of incalculable consequences to end such oppression' (cited in Ramsey, 1961, pp. 88–9).

So far as the just-war theory that the evil, damage, and suffering which war entails must be proportionate to the injury it is designed to avert or the injustice which occasions it, there is now general acceptance that the evil involved in the use of nuclear weapons would be wholly disproportionate to any good that could be achieved by using them. *The Church and the Bomb* (CEBSR, 1982, p. 144) states that their use would be an evil 'greater than any conceivable evil which the war is intended to prevent', while the *Pastoral Letter* of the US Catholic Bishops says that 'To destroy civilization as we know it by waging a "total war" as today it could be waged would be a monstrously disproportionate response to aggression on the part of any nation' (US Catholic Conference, 1983, p. 31).

A particular problem is raised by the concept of limited nuclear war, and by the possibility of using tactical nuclear weapons. For example, McMahan (1981, p. 121) has suggested that 'certain battlefield nuclear weapons, such as the neutron bomb, might be used in ways which satisfy [the principles of just-war theory]'. In its annual yearbook for 1982, however, the Stockholm International Peace Research Institute stated that 'the neutron bomb is not a prescription for a safe nuclear war for Europeans ... significant radiation casualties could be expected over an area of 10 square kilometres for each neutron weapon used: if 1,000 such weapons were used – and that is what might be needed – there could be anything up to some 10,000 square kilometres in which Europeans would be subjected to dangerous radiation exposure' (SIPRI, 1982, p. 10). Both the authors of *The Church and the Bomb* (CEBSR, 1982, p. 56) and the US Bishops' *Pastoral Letter* make it clear that they doubt whether a nuclear war could be controlled or limited: 'Today the possibilities for placing political and moral limits on nuclear war are so minimal that the moral task ... is prevention: as a people, we must refuse to legitimate the idea of nuclear war' (US Catholic Conference, 1983, p. 38). Indeed 'the burden of proof remains on those who assert that meaningful limitation is possible' (US Catholic Conference, 1983, p. 46).

The Principle of Non-combatant Immunity and Double Effects

The principle of non-combatant immunity is based on the belief that it is immoral for an aggressor to attack anyone who is not attacking, or intending to attack, him and on the more general moral principle that it is wrong to kill innocent people. In the context of a war an innocent person is one who is unarmed and not involved in the fighting.

Left at that, this principle, while not outlawing war as such, would make it virtually impossible to conduct any war justly since, in most actions, there is a high probability that some non-combatants will be killed. To get round this problem, the 'principle of double effects' is invoked. Under this principle, which can loosely be understood as 'the principle of side effects', it is argued that one can justly pursue a course of action which has a good end (in this case, trying to kill enemy combatants or damage their military effectiveness) even though one may possibly, probably, or even inevitably, cause unwanted harm by killing non-combatants. The reason is that the harm one causes to non-combatants is not *intended* in the strict sense of the word, and can therefore be regarded as incidental or accidental to one's main aim. One can not, however, invoke the principle of double effects to justify a course of action which would result in a disproportionate number of non-combatant deaths. In each case, the good and evil effects of the attack have to be weighed up, and if it is judged that the incidental harm which is likely to result is disproportionate to the good which is likely to arise from the attack, then the plan of attack must be abandoned. Similarly, if one could achieve one's objective in ways which would cause less harm to non-combatants, then one should do so.

The development of the concept of strategic or area bombing during the Second World War raised moral problems which were tackled by, for example, the American Catholic moralist, Father John Ford (1944). He subsequently extended his arguments to cover the atomic bombing of cities, arguing that those who planned or executed the hydrogen bombing of a city '(1) *would not* in practice avoid the direct intention of violence to the innocent; (2) *could not* if they would; and (3) even if they would and could avoid it, would have no proportionate justifying reason for permitting the evils which this type of all-out nuclear warfare would let loose' (Ford, 1957, p. 7). Both strategic bombing and the use of nuclear weapons in effect reverses the relationship between the good and evil effects of bombing since the military objective is in this case destroyed indirectly and incidentally as a part of the city, rather than vice versa. As Ford (1957, p. 8) put it: 'It is a question of its incidentality ... I doubt ... that the air strategist can drop his H-bomb ... and call the resultant death of millions of innocent people merely incidental.' He rejected the arguments of those who saw war as total: 'Beware of people who talk of modern war *as total*. It is total not because all the civilians wage it, but because all of them are its targets' (Ford, broadcast talk, in Ramsey, 1961, p. 71). This does not mean that they are legitimate targets.

The Universal Validity of the Principles of Just-War Theory

It is sometimes argued that humanitarian characteristics such as kindness, pity, compassion, mercy, toleration, etc. are universal in their application. Such characteristics are said to form the basis of the modern international law of armed conflict. Yet, as Best (1982, p. 162) points out,

> laws and rules placing too great a strain on the self-control, bravery, honour, self-sacrifice or goodwill of belligerents will simply not be observed when the going gets rough ... Having been commissioned to fight, [the warrior] insists he cannot do it with his hands tied behind his back. We must remark again that the law of war can never be more than a compromise between what men's humane desires make them wish to do, and what the release of their combative proclivities in war drives them to do ... real inescapable military necessity remains an irreducible fact of war ... setting limits to what humanitarianism can reasonably demand.

Best's position seems to infer that the most that can be said for just-war theory and for the international law of armed conflict is that it reflects what people would like to do, not what is necessarily done, and that as such it is normative (expressing standards to which we should strive) rather than prescriptive (laying down rules which must be obeyed) (p. 164). Best, who also points out that what is permissible is in part culturally determined, seems to deny that a universal ethic of war is possible (pp. 157–60). Nevertheless, he accepts that mutual adherence to the international law of armed conflict may be positively beneficial to both sides in a conflict because it tends to mitigate the amount of suffering arising from war; because in general it is beneficial to respect international treaties; and because violation of this international law may result in reprisals.

The Ethics of Deterrence

Utility

Arguments in favour of having nuclear weapons usually point to the utility of deterrence. Through its possession of nuclear weapons, the Western alliance is able to avoid the two evils of a conventional war and of almost certain defeat through such a war at the hands of the Soviet Union, with a consequential loss of freedom. This in no way implies a wish to use the nuclear weapons. Indeed, to do so would result in immeasurable damage and suffering for everyone involved in the conflict. However, it is partly the certainty of such damage that gives the nuclear deterrent its force. Of course, there are risks involved. If deterrence failed, and if the conflict could not be contained, then the results would be catastrophic, but so too would the results of a conventional war. Without a nuclear deterrent, the chances of a conventional war occurring would be much higher. Thus, while there are acknowledged dangers, possession of nuclear weapons has assured the continuing freedom of Western Europe and a prolonged period of peace.

These, it is said, are the positive benefits of deterrence, and they should not be undervalued. Those who argue against possession of the nuclear deterrent sometimes assert that it is 'self-evident that it is better to be red than dead. This is a bogus question; being red and being dead are not alternatives in the strict logical sense ... the intention of the proponent of deterrence, as much as that of the pacifist, is to remain neither red nor dead' (Second Permanent Under-Secretary of State at the Ministry of Defence, Arthur Hockaday, 1982, p. 80).

Morality

Many of those who argue in favour of the utility of deterrence accept that it is a bad way of coping with the threats facing the world. Hockaday (1982, p. 88) for example, approvingly cites Michael Walzer's statement that 'there may well be no other [way] that is practical in a world of sovereign and suspicious states. We threaten evil in order not to do it, and the doing of it would be so terrible that the threat seems in comparison to be morally defensible' (Walzer, 1978, p. 274). Walzer, however, goes further than this. He argues that

> nuclear war is and will remain morally unacceptable. Moreover, because it is unacceptable, we must seek out ways to prevent it, and because deterrence is a bad way, we must seek out others (p. 283).

Walzer admits that

> deterrence itself, for all its criminality, falls or may fall for the moment under the standard of necessity. But as with terror bombing, so here with the threat of terrorism: supreme emergency is never a stable position. The realm of necessity is subject to historical change. And, what is more important, we are under an obligation to seize upon opportunities of escape, even to take risks for the sake of such opportunities (p. 283).

One of the problems with deterrence to which Walzer alludes is its instability. Arms races, of which the present nuclear arms race is only the latest, have never been an infallible way of maintaining peace. Far from eliminating the causes of war, they tend to heighten tension and exacerbate international distrust. The 'peace' bought through deterrence, which is based on threat, distrust and fear, can be contrasted with 'genuine peace' based on international cooperation and understanding. The moral evil posed by the threat of nuclear war thus remains.

Pope John Paul II in his message to the Second UN Special Session on Disarmament (1982) argued that: 'In current conditions, *deterrence* based on balance, certainly not as an end in itself but as a step on the way toward a progressive disarmament, may still be judged morally acceptable. Nonetheless in order to ensure peace, it is indispensable not to be satisfied with this minimum, which is always susceptible to the real danger of explosion' (John Paul II, 1982, p. iii).

In their *Pastoral Letter* the US Catholic Bishops argued that 'the moral duty today is to prevent nuclear war from ever occurring *and* to protect and preserve those key values of justice, freedom and independence which are necessary for personal dignity and national integrity' (US Catholic Conference, 1983, p. 50). On this basis, they accepted Pope John Paul II's judgement that deterrence may still be judged to be morally acceptable, but they could not consider it 'adequate as a long-term basis for peace' (pp. 53–4). This led them to enunciate certain criteria for morally assessing the elements of deterrence policy. These include rejecting any plans that tend to encourage the view that nuclear wars could be fought or are winnable; rejecting the quest for nuclear superiority; and actively pursuing steps towards disarmament (p. 54). Although the Bishops specifically refused to condemn all aspects of nuclear deterrence they made it clear that 'the lack of an unequivocal condemnation of deterrence is meant only to be an attempt to acknowledge the role attributed to deterrence, but not to support its extension beyond the limited purpose ...[of]... prevent[ing] the *use* of nuclear weapons by others.' Thus 'any claims by any government that it is pursuing a morally acceptable policy of deterrence must be scrutinised with the greatest care' (p. 56).

Is it Wrong to Intend to do that which is Morally Wrong?

Although deterrence does not involve an active desire to use the nuclear weapons which a country has at its disposal, it does depend for its credibility on convincing the potential aggressor that one would be willing to use them should this be necessary.

Normally one thinks that if an action is immoral, then the intention to carry it out is also immoral. This is why conspiracy is regarded as a crime. However, G.R. Dunstan, Professor of Moral and Social Theology at King's College, University of London, has argued that a *conditional* intention to use immoral means is not necessarily immoral.

> The intention in possessing a nuclear 'deterrent' force is *not* to use it, but to restrain a potential enemy from a first, provocative use. The *intention* is, by maintaining a credible threat, to prevent any occasion for its use – to deter the other side from the first, immoral, act, the nuclear strike. If, knowing the consequences, he commits that act, the responsibility for the consequence is primarily his, however much the respondent also is to blame: he brings indiscriminate destruction upon his own head, and all his people. While he is deterred from evil – a nuclear attack – by the threat of nuclear retaliation, the maintenance of that threat could be morally justified. To deny justification for carrying out that threat would be to rob the deterrent of its force (Dunstan, 1982, p. 50).

Dunstan is at pains to point out that he is not asserting that

> to hold the deterrent, or to use it, is a 'Christian' act. The problem is one of those tragic necessities which ... cannot be categorised at all in Christian terms. There is no Christian solution to it. There is only a choice among evils; and there is the Everlasting Mercy of those who, in good faith, are driven to choose (p. 50).

It is important to note that Dunstan is *not* endorsing the NATO policy of first-use of nuclear weapons which has been condemned by, among others, the US Catholic Bishops. He is only talking about a retaliatory strike.

Hockaday takes a slightly different line in his defence of deterrence. He argues that

> it seems reasonable to posit a moral distinction between a simple intention x and a conditional intention 'if A, x' when a principal objective of forming and declaring the conditional intention is to secure, so far as lies in one's power, that A shall not come to pass and that the condition shall not be fulfilled (Hockaday, 1982, p. 84).

In saying this, Hockaday is careful to point out that he is

> not seeking to argue that the conditional intention is so morally laudable or morally neutral as to involve no element of evil. What I am arguing is that, although the conditional intention may contain an element of moral evil, a strategy of deterrence involving the conditional intention may be the most effective way of securing the

> twin objectives of preventing war and checking political aggression and may therefore be a morally acceptable price to pay to achieve those objectives (p. 84).

Thus

> although to have such a conditional intention may not be *good*, and may even involve elements of evil, I suggest that it may be *right* if, by virtue of being in the circumstances the most likely means of securing that peace shall be preserved and that nuclear weapons shall not be used either by myself or by others, it is the least evil of a number of prima facie evil courses that may lie before me (p. 85).

Thus far, the arguments put forward by Dunstan and Hockaday are fairly similar. Dunstan's argument, however, seems to ignore the fact that the leader of a country which has already been devastated by a nuclear attack still has a moral choice: to use the deterrent in retaliation, or not. Plainly, once one's country has been attacked, the basic rationale for the deterrent (to deter such an attack) has gone. The choice is then between refraining from oneself committing evil, or adding to the amount of evil that has already been enacted by using one's nuclear weapons in a retaliatory strike. Hockaday attempts to meet this point by arguing (1982, p. 83) that, for a deterrent to maintain its credibility it requires that 'when I seek to deter by expressing the conditional intention 'if *A*, *x*', I must not have formed an actual intention not to do *x*. I do not have to form an actual intention to do *x* until the condition *A* is satisfied.' Thus the decision to use the deterrent in retaliation is still open after the attack has taken place.

The Morality of Retaliation

D.H. Hodgson (1967, pp. 85–90) has argued that, for the deterrent theorist, the best outcome is for neither side to suffer nuclear devastation; the next best is for only one side (preferably not one's own) to do so; and the worst is for both sides to be devastated. Assuming that a nuclear war has broken out and that the best outcome of deterrence is no longer available, the rational choice is to opt for the second-best outcome. In these circumstances the leader of a country which has been devastated is debarred from retaliating because that would result in the worst outcome (both sides being devastated). This being so, the credibility of deterrence is wholly undermined. Thus there is a paradox at the heart of deterrence theory which revolves around the utility of the outcomes once deterrence fails.

Gregory Kavka (1978) has pointed to another paradox which centres on the morality of the agent who has to decide whether or not to retaliate. Kavka starts by arguing that, in general, the most useful act, which he defines as the one having the highest expected utility, is the one which should be performed whenever a very great deal of utility is at stake. He then looks at the case of deterrence. In this case, he says, '(i) a great deal of *negative* utility is at stake ... (ii) people will be likely to suffer serious injustice if the agent fails to perform the most useful act ... (and) (iii) performing the most useful act involves, at most, a small *risk* of performing an inherently evil act' (Kavka, 1978, p. 288). In the case of deterrence, the great deal of negative utility that is at stake is nuclear war; people will suffer if the agent (and deterrent) fails; and there is little risk that the agent will in fact have to use the deterrent, and thus commit an evil act.

Kavka then considers deterrence in the light of the 'Wrongful Intentions Principle ... To intend to do what one knows to be wrong is itself wrong' (p. 289). In line with Dunstan and Hockaday, he accepts that 'while the object of his deterrent intention might be an evil act, it does not follow that, in desiring to adopt that intention, he desires to do evil, either as an end or as a means' (p. 291). Thus Kavka believes that the moral agent can accept deterrence since the desire to have the intention to do an act *as a means of deterrence* is entirely consistent with having a strong desire not to do the act.

What happens, though, if deterrence fails? Paradoxically, Kavka argues, the *moral* agent is a 'captive in the prison of his own virtue' (p. 291). Although a rational moral agent 'should want to form the conditional intention to apply the sanction if the offence is committed, in order to deter the offence' (p. 291), this leaves open the question of whether or not he will apply it. Kavka believes that there would be 'conclusive moral reasons not to apply the sanction if the offence were committed (p. 292). A moral agent who recognizes that there are conclusive moral reasons for not applying the sanctions but chooses to ignore them 'is not moral' (p.

292). The only way a moral agent could bring himself to do something which he recognizes there are conclusive moral reasons not to do would be to corrupt himself (p. 295). But, says Kavka, 'we regard the promotion and preservation of one's own virtue as a vital responsibility of each moral agent, and self-corruption as among the vilest of enterprises' (p. 297). In other words 'we assent to the Virtue Preservation Principle: It is wrong to deliberately lose (or reduce the degree of) one's moral virtue' (p. 298). Yet this is precisely what the moral agent who decides to use the deterrent would do. The moral agent thus has conclusive moral reasons for not using the deterrent.

Conclusions

The development of nuclear weapons has raised major moral problems which philosophers and theologians have had to face. As we have seen, there is no consensus on the issues although there is wide agreement that the first-use of nuclear weapons would be an evil, and that their use would breach the just-war principles of proportionality and non-combatant immunity. There are also good reasons why, on utilitarian grounds, they should not be used. Many philosophers, however, have conceded, again on utilitarian grounds, that nuclear deterrence has brought some benefits (of peace and security) but others have argued that the peace is fragile and the sense of security is based on fear. More importantly, philosophers have begun to expose the essential paradox of deterrence by pointing out that the failure of deterrence would not remove moral choice from the act of retaliation. The use of nuclear weapons in retaliation would add to the evil already done, so that the virtuous man should refuse to use them in retaliation.

References and Further Reading

Allen, R.V. (1981) cited in *The Times*, 23 March.

Best, G. (1982) 'International Humanitarian Law: Principles and Practice', in G. Goodwin (ed.) (1982) *Ethics and Nuclear Deterrence*, London: Croom Helm.

CEBSR (Church of England Board for Social Responsibility) (1982) *The Church and the Bomb. Nuclear Weapons and Christian Conscience*, London: Hodder and Stoughton.

Dun, A. and Niebuhr, R. (1955) 'God Wills Both Justice and Peace', *Christianity and Crisis*, 15, 10.

Dunstan, G.R. (1982) 'Theological Method in Deterrence Debate', in G. Goodwin (ed.) (1982) *Ethics and Nuclear Deterrence*, London: Croom Helm.

Ford, J.C. (1944) 'The Morality of Obliteration Bombing', *Theological Studies*, 5 (3), 261–309.

Ford, J.C. (1957) 'The Hydrogen Bombing of Cities', *Theology Digest*, Winter. Reprinted in W.J. Nagle (ed.) (1960) *Morality and Modern Warfare*, Baltimore: Helicon Press.

Glover, J. (1977) *Causing Death and Saving Lives*, Harmondsworth: Penguin Books.

Harris, J. (1980) *Violence and Responsibility*, London: Routledge and Kegan Paul.

Hockaday, A. (1982) 'In defence of deterrence', in G. Goodwin (ed.) (1982) *Ethics and Nuclear Deterrence*, London: Croom Helm.

Hodgson, D.H. (1967) *Consequences of Utilitarianism. A Study of Normative Ethics and Legal Theory*, Oxford: Oxford University Press.

Kavka, G.S. (1978) 'Some paradoxes of deterrence', *The Journal of Philosophy*, 75 (6), 285–303.

John XXIII (1967) 1st edn., *Pacem in Terris*, London: Catholic Truth Society.

John Paul II (1981) 'Address to Scientists and Scolars', *Origins*, 10, cited in United States Catholic Conference, Inc. (1983) *The Challenge of Peace: God's Promise and Our Response,* London: Incorporated Catholic Truth Society and SPCK.

John Paul II (1982) Message to the Second Special Session of the United Nations General Assembly Devoted to Disarmament, June 1982, cited in United States Catholic Conference, Inc. (1983) *The Challenge of Peace: God's Promise and Our Response,* London: Incorporated Catholic Truth Society and SPCK.

Kent, B. (1982) 'A Christian Utilitarianism from a Christian Background', in G. Goodwin (ed.) (1982) *Ethics and Nuclear Deterrence*, London: Croom Helm.

Lipscombe, O. (1982) in *The New York Times*, 19 November.

McMahan, J. (1981) *British Nuclear Weapons: For and Against*, London: Junction Books.

Pius XII (1944) Christmas Message, cited in Ramsey, P. (1961) *War and the Christian Conscience. How shall modern war be conducted justly?* Durham, North Carolina: Duke University Press, 84.

Ramsey, P. (1961) *War and the Christian Conscience. How shall modern war be conducted justly?* Durham, North Carolina: Duke University Press.

Regan, T. (1972) 'A Defense of Pacifism', *Canadian Journal of Philosophy*. 11 (1), 73–86.

Singer, P. (1979) *Practical Ethics*, Cambridge: Cambridge University Press.

SIPRI (Stockholm International Peace Research Institute) (1982) *The Arms Race and Arms Control*, London: Taylor and Francis Ltd.

United States Catholic Conference (1983) *The Challenge of Peace: God's Promise and Our Response. The US Bishop's Pastoral Letter on War and Peace in the Nuclear Age*, London: Incorporated Catholic Truth Society and SPCK.

Walzer, M. (1978) *Just and Unjust Wars*, London: Allen Lane.

10
Disarmament, Arms Control and Non-Nuclear Defence

The Rejection of Disarmament, 1949–55

Following the failure of the Baruch Plan to prohibit the manufacture of atomic weapons and plan the development and use of atomic energy under an International Commission and the Soviet counterproposals (see pp. 8–9) there was a brief and final attempt to negotiate the total prohibition of atomic weapons in the early 1950s. These negotiations took place in the UN Disarmament Commission following an Anglo-French initiative in June 1952, subsequently taken up by the USA, which linked proposals for reductions in conventional force levels with the renunciation of nuclear weapons as offensive weapons pending their total prohibition. Considerable progress had been made by May 1955 when the Soviet Union responded with concrete proposals including, in respect of nuclear weapons, plans for their prohibition and the acceptance of a system of UN inspections as a means of verifying compliance. The West, however, having at first 'been gratified to find that the concepts which we have put forward over a considerable length of time… have been accepted in large measure by the Soviet Union' (US delegate to the UN Disarmament Commission, 12 May 1955, cited in Philip Noel-Baker, 1958, p. 22), insisted on a recess. In the next few weeks opposition to the proposed treaty grew. The US Atomic Energy Commission argued that complete elimination of nuclear weapons was impossible because stocks were so high that it would be impossible to account completely for all past production of nuclear materials. The USA therefore queried the feasibility of the international control mechanisms that had been suggested and put forward proposals for aerial reconnaissance (Eisenhower's 'Open Skies' Policy) which were known to be unacceptable to the USSR. When the UN Disarmament Commission resumed its meeting the Soviet Union tried to steer the discussions back to the point that had been reached in May, whereupon the US delegation withdrew their original proposals.

In effect, by the mid-1950s policy makers in the USA had rejected the idea of banning nuclear weapons (nuclear *disarmament*) and were seeking instead ways of controlling nuclear weapons (*arms control*), thus strengthening deterrence with a view to reducing the risk of nuclear war. While some saw arms control as a first step towards disarmament, most people came to regard it as a means of making life with the bomb bearable. Reluctantly they came to accept nuclear deterrence.

Proliferation

In parallel with the efforts to control nuclear arms there were efforts to control their spread to other countries. From the early 1940s it was recognized that nuclear technology could be used to make weapons and to generate power. Although the early work on the atom bomb was done in Britain the British transferred the technology to the USA where resources were greater. British scientists nevertheless worked on the Manhatten Project until in 1946, anxious to protect its monopoly, the USA stopped further collaboration with the passage of the McMahon Act (see chapter 1).

By the early 1950s, with three nuclear-weapon states (the USA, USSR and Britain), it was clear that a policy of secrecy would not prevent the spread of nuclear weapons. At the same time there was increased interest in the peaceful exploitation of nuclear technology for power generation, thus providing export possibilities for the West's nuclear industries while coupling this with control on its use for weapons programmes. In December 1953 the US announced its 'Atoms for Peace' programme under which it promised to help other countries develop nuclear power programmes. Initially the Americans accepted a paper declaration that the technology they were exporting would only be used for peaceful purposes. Subsequently they turned to the idea of on-site inspection, a function which was vested in the International Atomic Energy Authority (IAEA), a UN agency set up in 1957 to promote the peaceful development of nuclear applications and to establish and administer an international system of safeguards. The US also organized (in 1954) a secret cartel of uranium suppliers (the Western Suppliers Group) whose members regularly exchanged information on uranium sales.

Notwithstanding these moves, the French secretly agreed to provide Israel with a reactor capable of producing plutonium (1957) and the Israelis also acquired the blueprints for a reprocessing plant from a French company. By 1968 US intelligence believed that Israel already had nuclear weapons. By then France and China were also nuclear-weapon states while both West Germany and Japan had the capability to develop nuclear weapons. US concern about proliferation was reflected in the report of a Special Presidential Commission on Nuclear Proliferation (Gilpatric Committee, 1965, pp. 2–3) which concluded that 'the spread of nuclear weapons poses an increasingly grave threat to the security of the United States.' As a result the US proposed a Non-Proliferation Treaty to the Eighteen Nation Disarmament Conference, agreement on which was finally reached in 1967.

The Non-Proliferation Treaty (NPT) effectively confirmed a club of nuclear-weapon states (those nations which had tested a nuclear weapon before 1 January 1967) and set out to prevent the transfer of nuclear weapons or nuclear explosive devices from the nuclear-weapon states to non-nuclear-weapon states. It does not however affect the 'inalienable right of all the Parties to the Treaty to develop research, production and use of nuclear energy for peaceful purposes'. Non-nuclear-weapon states agreed to accept IAEA safeguards which, at the time the NPT opened for signature, had still to be defined. The Treaty also allows states to withdraw from it on 90-days notice. Subsequently the three depository states (USA, USSR and Britain) promised immediate assistance to a non-nuclear-weapon state party to the NPT if it became the 'victim of an act or an object of a threat of aggression in which nuclear weapons are used' (UN Security Council Resolution 255, 1968). Critics of the Treaty argue that this guarantee is worthless.

International safeguards were finally agreed by the IAEA Board of Governors in 1971 (IAEA document INFCIRC/153). These call, firstly, for a national system of accounting for and control of nuclear material to ensure that none is diverted from peaceful uses to nuclear weapons use, including an inventory of fissile material coupled with independent means of verification through measurement and observation; secondly, for provision of information on nuclear material and facilities to the IAEA; thirdly, for inspection of nuclear facilities by IAEA inspectors; and fourthly, for IAEA safeguards on all transfers of nuclear materials.

The NPT has not, however, prevented proliferation. The making of nuclear weapons is now neither enormously expensive nor beyond the capability of many countries' scientific communities. On-site inspection cannot prevent states determined to acquire nuclear weapons from doing so, as a leaked IAEA document COV/1842.8 (1977) admitted. Since the IAEA has to give notice of an inspection, material can be temporarily returned to installations about to be inspected by the IAEA; records can be inadequate; and audit visits can be delayed or impeded. It is in any case impossible to predict precisely how much fissile material will be produced by a particular facility or lost from the system as material unaccounted for (MUF). Since the amounts needed to make a bomb are small (4–5 kg) and well within the margin of error of larger nuclear programmes, the inspection system cannot prevent the spread of nuclear weapons. The fact that it takes only about ten days to make a bomb means that material could be diverted and made into bombs between IAEA inspections. A further problem is that although 118 countries had signed the NPT by 1983, France, China, India, South Africa, Pakistan, Spain, Argentina, and Brazil, all of whom have significant nuclear programmes, have not signed it.

In May 1974 India exploded a small 150-kiloton bomb using material obtained from its Canadian supplied research reactor. Following this event the Canadians proposed reconvening the Western Suppliers Group to control the spread of nuclear technology. At a secret meeting in London the USA, Britain, France, Canada, West Germany, Japan, and the USSR drew up a list of sensitive technology which they agreed would not be exported unless proper safeguards on its use were agreed by the recipients (the 'trigger list'). The group, whose existence was acknowledged in 1975, was subsequently enlarged and is known as the London or Nuclear Suppliers Group. However, its members have not always acted in concert. Under President Carter the USA sought to place an embargo on the export of reprocessing and enrichment plants (both of which are required to obtain bomb-grade plutonium), and to place tougher safeguards on the export of all materials and equipment supplied by the USA (US Non-Proliferation Act). Some of America's allies refused to impose similar safeguards which they felt would inhibit their exports. Indeed the Nuclear Suppliers Group has consistently voted down proposals to make 'full-scope' safeguards a condition of nuclear sales and non-signers of the NPT are therefore still able to import much of their normal nuclear supplies from the Suppliers.

The NPT has failed to control proliferation: there is evidence that the South Africans exploded a small nuclear device in the South Atlantic (September 1979) (*The Guardian*, 22 May 1985) while Pakistan is said to 'have what they need to produce their own nuclear weapons' (*The Guardian*, 22 February 1984) and may have only refrained from testing a bomb because the Chinese have vetted their design (*The Guardian*, 31 October 1984). It has also been alleged that Argentina has enough plutonium to make about ten bombs and that Brazil may soon be able to produce nuclear weapons.

Similar in nature to the Non-Proliferation Treaty is the Treaty for the Prohibition of Nuclear Weapons in Latin America (Treaty of Tlatelolco) which came into force in 1967. This prohibits the possession, manufacture or testing of nuclear weapons by parties to the treaty. The treaty calls for full-scope safeguards by the IAEA. The importance of the Treaty of Tlatelolco is that it is a model for the creation of similar nuclear-weapon-free zones elsewhere. However, while 22 states have acceded to it, Argentina (signed but not ratified), Cuba and Guyana have not as yet done so, and the treaty is also not yet in force in Brazil or Chile on the grounds that it has yet to be ratified by all the states in the zone. Cuba's reluctance to sign the treaty is related to controversy over the US base at Guantanamo.

The overall situation is not encouraging, particularly given the pressures to 'go nuclear' that stem from the belief that small independent nuclear deterrent forces are sensible and that nuclear weapons status confers prestige on the nation, and from the internal pressures generated by political, military and scientific-industrial interests to develop nuclear weapons production facilities. There are some doubts that the NPT will be renewed when it next comes up for review in 1985.

Nuclear Arms Control and Confidence-Building Agreements

While disarmament negotiations have concentrated on the abolition of weapons, usually through phased but significant reductions, the overriding purpose of arms control has been to enhance security through agreements which fall short of the abolition of nuclear weapons. There are two main groups of arms-control agreements: *multilateral* ones that are open to accession by all states, and *bilateral* ones, most of which are between the Soviet Union and the USA.

Multilateral Agreements

Multilateral nuclear arms-control agreements have in general been concerned with preventing the introduction of nuclear weapons in specific environments and controlling the testing of nuclear weapons. The first category includes the 1967 Treaty on Principles Governing the Activities of States in the Exploration and Use of Outer Space including the Moon and other celestial bodies (the Outer Space Treaty), and the 1972 Treaty on the Prohibition of the Emplacement of Nuclear Weapons and other Weapons of Mass Destruction on the Sea-Bed and the Ocean Floor and in the Sub-soil Thereof (the Sea-Bed Treaty). The former prohibits the testing or placement of nuclear weapons in orbit around the earth, installing them on celestial bodies, or stationing them in outer space in any other manner, and has been acceded to by 81 states including all NATO and Warsaw Pact states. The latter prohibits the deployment of nuclear weapons or related facilities on the bottom of the sea for any purposes. 70 states, including all NATO and Warsaw Pact littoral states have acceded to the Treaty. Finally, the Antarctic Treaty (1959) prohibits nuclear bases, nuclear explosions or the disposal of radioactive waste, in Antarctica.

At the time that it was concluded, the Outer Space Treaty concerned an area of only limited military usefulness. Unfortunately, advances in space technology and the relatively narrow wording of the Treaty means that outer space has not been denuclearized. The Treaty does not prevent the use of so-called fractional orbital bombardment systems (FOBS) which are nuclear weapons placed in orbit but flying less than one full revolution round the Earth (although these were prohibited under the SALT II agreement). Nor does it prevent the flight through outer space of ballistic missiles carrying nuclear weapons. More importantly, the deployment in outer space of weapons not capable of mass destruction is unrestricted. Thus killer satellites and BMD interceptors (including charged-particle beam weapons but excluding those operating with nuclear warheads) are not prohibited. The most immediate problem is the destabilizing impact of ASAT developments.

The Sea-Bed Treaty does not forbid the militarization of the Sea-Bed, only the emplacement of nuclear weapons on it or within its subsoil and the deployment of bottom-crawling vehicles specifically designed to carry nuclear weapons. It does not prevent the temporary stationing of SSBNs on the seabed and it does not cover the seabed inside the 12-mile limit. In general, the military significance of the treaty at present is slight. Dan Smith (1980, p. 218) notes that 'it is now common to liken this treaty to an agreement in which states agree not to bolt their aircraft to the ground'.

In the second category of controlling tests of nuclear weapons, there is the Partial Test Ban Treaty (1963) under which the USA, the Soviet Union and Britain agreed to ban nuclear weapons tests in the atmosphere, in outer space and under water or in any environment if the explosion would cause radioactive debris to affect areas outside the territorial limits of the state doing the testing. The Treaty is open to all states for signature and 111 have acceded to it. However, it has not been signed by the People's Republic of China or France, both of which have continued atmospheric testing (although France ceased in 1974). While the Treaty has helped to curb radioactive pollution, it has not stopped underground nuclear testing. A subsequent bilateral Treaty between the USA and the Soviet Union on the Limitation of Underground Nuclear Weapons Tests (the 1974 Threshold Test Ban Treaty) prohibited underground nuclear weapons tests carried out for a military purpose and having a yield exceeding 150 kilotons from 31 March 1976. This latter treaty did not cover underground nuclear explosions carried out for peaceful purposes. These were covered by the 1976 US–Soviet Treaty on Underground Nuclear Explosions for Peaceful Purposes (the Peaceful Nuclear Explosions Treaty) which sought to limit the yield of individual peaceful nuclear explosions to no more than 150 kilotons and to ensure that they could not be used for military purposes. Neither the Threshold Test Ban Treaty nor the Peaceful Explosions Treaty have entered into force, although their provisions appear to have been observed.

As an arms control measure the Partial Test Ban Treaty was 'utterly useless and positively harmful precisely because it did calm public opinion' (Smith, 1980, p. 217). So far as the Threshold Test Ban Treaty is concerned, there is widespread agreement that the 150-kiloton limit on yield is much too high and does not represent a meaningful limitation on the nuclear-weapon states (Frei, 1983, p. 185).

Bilateral Agreements

Reference has already been made to two bilateral nuclear arms-control treaties – the Threshold Test Ban Treaty and the Peaceful Nuclear Explosions Treaty, both of which are concerned with the testing of nuclear weapons. Other bilateral agreements have been concerned with confidence building measures and with the establishment of direct communication links aimed at providing emergency communications during times of crisis. Among the latter is the original US–Soviet 'Hot line' agreement of June 1963 and the 'Hot line' modernization agreement of 1971; the Soviet–British direct communication link agreement of 1967; the US–Soviet Agreement on Measures to Reduce the Risk of Outbreak of Nuclear War between the USA and the USSR (1971), which requires immediate notification of and a commitment to take preventative action to prevent any unauthorized detonation of a nuclear weapon, and immediate notification of the detection of an unidentified object by missile warning systems or of interference with the warning system; and advance notification of planned missile launches beyond borders in the direction of the other party. (Analogous agreements between France and the Soviet Union and Britain and the Soviet Union were signed in 1976 and 1977 respectively). A further US–Soviet Agreement on the Prevention of Nuclear War (1973) enjoins each party to refrain from the threat or use of force against the other party or its allies in circumstances endangering international peace and to hold urgent consultations in situations involving risk of nuclear war. The most important bilateral agreements have been the SALT I and SALT II Agreements and the ABM Treaty, the provisions of which were discussed in chapter 2 (pp. 19–21).

The SALT I Treaty was important in as much as it contributed to a relaxation of tension between the USA and the Soviet Union and in general advanced detente. It did not, however, slow down the arms race and it may in fact have helped accelerate it in new directions, particularly into MIRV technology. Asymmetries were dealt with by 'upward arms control', that is, broadly by allowing the disadvantaged side to catch up with the other side, and by a process of managing rather than terminating the existing arms race. Alva Myrdal (1980, p. 127) commented that 'By no stretch of the imagination can this be called arms limitation. Instead it is a mutually agreed continuation of the arms race, regulated and institutionalized. The competition for quality of nuclear weapons remains totally unregulated, leaving open the avenue for gaming without end.'

Some of these drawbacks were taken up in SALT II which not only set quantitative limits but also qualitative limits by prohibiting several technologically feasible undertakings. On the other hand, these limitations were only temporary and the arms race was not ended. In particular, the overall ceilings and sub-ceilings established by SALT II were deliberately set high, allowing both superpowers to continue their exploitation of MIRV technology.

Generally speaking the 1972 ABM Treaty is considered to be the most successful arms-control agreement of the 1970s because it helped to prevent a competition between ABM systems and technologies designed to penetrate them. Even Myrdal, a leading critic of the effectiveness of arms control, concedes that the treaty was relatively effective in inhibiting development of the arms race into a new area of technology. Its most significant failing was that it allowed continued testing of ABM components at designated test ranges. The USA argues that its present Ballistic Missile Defence (BMD) research and development programme is not only within the ABM Treaty but also, by keeping the USA abreast of BMD technologies, actively discourages the Soviet Union from abrogating the Treaty. The Soviet Union is also researching in the field. However, the fact that such R & D takes place encourages both sides to continue their work and, in effect, constitutes an R & D arms race and raises fears that one or other will reach a position in which it feels it would be advantageous to abrogate the ABM Treaty and begin to deploy 'star wars' type defences.

Treaty Violation and Verification of Compliance

In spite of the general assumption that international treaties will be carried out in good faith, it is accepted practice to supervise performance. Verification of treaty compliance has assumed new importance with the advent of arms-control agreements. Such agreements are 'policed' in a number of ways: for example, national technical means of verification encompass reconnaissance satellites and aircraft and ship- and shore-based listening posts and are designed to verify compliance without direct cooperation from the state under surveillance. The reliability of such means is disputed. The former US Secretary of State, Harold Brown, when asked whether the USA would know whether the Soviet Union was trying to conceal activities or impede verification, told the Senate Committee on Foreign Relations that 'We will be able to see when that happens. We will not always be sure, but we will always know enough to be able to raise it as an issue... It is easier to tell that they are trying to hide something than it is to tell what they are trying to hide' (Senate Committee on Foreign Relations, 1979). In 1972 the Soviet Union and the USA established a Standing Consultative Committee to promote the implementation of the Nuclear Accidents Agreement, the SALT ABM Treaty and the SALT I Agreement and to discuss any charges by either side of violations of the SALT accord. Initially where the USA raised issues related to treaty compliance with the Soviet Union, the queries tended to be of marginal significance and their resolution showed that the treaties were being observed.

The Standing Consultative Committee has continued in being in spite of the poor relations between the superpowers. Charges of violation have, however, become more frequent and early in 1984 both the USA and the Soviet Union drew up comprehensive lists of their complaints against each other and made the lists public. Soviet charges of US infringements have generally related to the spirit rather than the letter of the agreements, and as such are not generally capable of substantiation on a strict reading of the treaties. Conversely, the charges brought by the USA against the Soviet Union have mostly been vague and conjectural and most of them must be regarded as shallow propaganda (Goldblat, 1984, p. 185). The USA has nevertheless been taking a more serious view of alleged Soviet violations of late.

The greatest danger is that one or the other superpower will secretly develop a super-weapon that will give it a clear superiority over the other. However, Congressman Les Aspin, for example, has argued that

> The introduction of a new strategic weapon involves at least five stages: research, development, testing, production and deployment. At any one of these stages the present ability of the US to detect clandestine activity on the part of the USSR ranges from fair to excellent. The key point, however, is that the Russians would have to disguise all five stages, and that the odds against their successfully doing so are extremely high (Aspin, 1979, p. 30).

Harold Brown estimated that the chance of the Soviet Union being able to develop a weapon system without the USA being aware of the fact was 'substantially less than 1 out of 1 million' (Senate Committee on Foreign Relations, 1979).

Not everyone agrees that problems of verification have been solved. For example, Moreton (1983, p. 55) argues that while 'a ban on flight-testing might be quite easy to monitor... neither a comprehensive test ban, nor a halt on the production of weapons-grade fissile materials can as yet be verified with sufficient certainty to ensure that a real freeze is being observed. The stakes are too high to allow much room for ambiguity.' Frei (1983, p. 30) notes that both the US and Soviet governments 'are aware that the traditional verification procedures, especially the "verification by national means" (i.e. by satellite observation and conventional intelligence), although constantly being improved, may, in the foreseeable future, become insufficient and lag behind the increasing complexity and sophistication of the strategic weapons systems.' Frei (1983, p. 30) notes the concern expressed by Georgi Arbatov (1981, p. 181) that the characteristics of, for example, cruise missiles and mobile ballistic missiles complicate arms control and also notes the comment of Eugene Rostow, Director of the US Arms Control and Disarmament Agency, that 'verification has become a more and more troublesome aspect of arms control in recent years.'

Arms Control: Success or Failure?

Generally speaking, arms control has not been noted for its success. Although Moreton (1983) argues that 'the list of arms control agreements signed by the two superpowers... over the past two decades is quite impressive' (p. 40) she accepts that 'from a vantage point in the 1980s, the arms control record of the 1970s is disappointing' (p. 34). In particular they 'have put only the loosest of limits on the nuclear arsenals of both sides'. Dahlitz (1983, p. 17) points out that 'the main thrust of the [multilateral] treaties is to prevent the spread of nuclear weapons and to preserve the status quo, but no attempt is made to dismantle existing nuclear arms and facilities.' In this sense they are largely concerned with the problem of 'horizontal proliferation', that is with the spread of nuclear weapons to non-nuclear-weapon states. The various bilateral treaties, on the other hand, seek to curtail quantitative increases in and, to a lesser extent, qualitative improvements of, nuclear weapon systems between the existing nuclear powers (notably the two superpowers). They have therefore been more concerned with the problem of 'vertical proliferation'. Few observers believe that they have succeeded. Leitenberg (1978, p. 20) concludes that 'The meagre "partial" arms control measures have been futile, ineffective and outside the main thrust of great Power military efforts.' Epstein (1976, p. 36) holds that the agreements are designed 'not to halt or reverse the arms race but rather to institutionalize it and regulate it so that it may continue within each country on its own momentum and under conditions of relatively less instability and insecurity for the two great powers – in other words, a blueprint for the continuation of the arms race under agreed-upon terms and conditions.'

Arms control is largely concerned with stability. This is not, however, necessarily served by reductions in strategic arms but rather in the survivability of a retaliatory force. In some circumstances this may actually call for an *increase* in nuclear weapons. Security in this sense depends on minimizing the vulnerability of one's own strategic retaliatory forces (e.g. through mobile-basing on land, SSBMs, hardening of silos, etc.). It can be undermined if the other side tries to develop counter-force weapons that threaten one's strategic forces *before* they are fired (e.g. highly accurate missiles, ASW capabilities) or destroys them once they have been fired (ballistic missile defence systems). It is for this reason that the present US arms programme is destabilizing.

Reviewing the impasse in arms-control talks in 1984, Harold Brown and Lynn Davis (1984, p. 1146) have pointed out that 'Negotiations with the Soviet Union cannot be expected to produce in the foreseeable future nuclear force postures with fewer than many thousands of nuclear weapons. Nuclear war will still be possible and its catastrophic consequences will not be changed.' In addition, they state 'negotiations would have to produce fairly significant reductions or limits on new weapons to arrest the drift towards more vulnerable nuclear forces and toward greater instabilities as a consequence of the introduction of new technologies' (p. 1146), and 'it also remains difficult to design arms control proposals that appear to be equitable' given the different size and characteristics of US and Soviet nuclear force (p. 1146).

Arms Control and Disarmament Proposals

The failure of arms control to stem the arms race or indeed to come to any concrete agreements since the (unratified) SALT II agreement, coupled with increased public concern, precipitated a major public debate in both the USA and in Europe on various means of curbing or reversing the arms race.

A Comprehensive Test Ban Treaty (CTBT)

Although the 1963 Partial Test Ban Treaty succeeded in banning atmospheric nuclear explosions and reducing the level of radioactive fall-out, the nuclear-weapon states have continued to test their weapons underground. The testing is mainly in order to find ways of increasing their efficiency and to develop new weapon designs; to develop mechanisms to ensure the safety and security of nuclear devices; to evaluate the effects of nuclear explosions; and to maintain confidence in the reliability of stockpiled weapons. The peace movement argues that a ban on testing would be an important step towards preventing the development of new and more sophisticated nuclear weapons and would raise the nuclear threshold by introducing an element of uncertainty about the reliability of warheads in the stockpile. On the other hand, it is argued that qualitatively new developments in nuclear weapons technology are unlikely and that even a ban on testing would not prevent *all* improvements although it would make it more difficult if not impossible for nuclear-weapon states to develop new designs, and would constrain the modification of existing designs. It has also been suggested that the continued operability of stockpiled nuclear weapons can be achieved by non-nuclear testing (SIPRI, 1982, p. 128).

Proposals for a CTBT have been under discussion for many years but a number of issues, notably the question of verification, have not been resolved by the superpowers even though as long ago as 1972 the Secretary General of the UN stated that 'all the technical and scientific aspects of the problem have been so fully explored that only a political decision is now necessary in order to achieve final agreement' (SIPRI, 1983, p. 232). Since then the techniques of verification have improved.

The USA in particular has been reluctant to negotiate a CTBT on the grounds that such a ban could not help 'to reduce the threat of nuclear weapons or to maintain the stability of the nuclear balance' (Committee on Disarmament document CD/PV. 152), and it has relegated a CTBT to being a 'long-term goal' rather than the 'high-priority objective' which it is for most countries. The USA also appears to believe that continued testing is necessary in its own right. Replying to questions from the House Appropriations Committee, the US Arms Control and Disarmament Agency said that 'nuclear tests are specifically required for the development, modernisation, and certification of nuclear warheads, the maintenance of stockpile reliability, and the evaluation of nuclear weapon effects' (*The Guardian*, 9 August 1983). At present it seems unlikely that a CTBT would be acceptable to the USA, Britain, France or China.

A Freeze

In April 1980 Randall Forsberg, Director of the US-based Institute for Defense and Disarmament Studies, issued a 'call to halt the nuclear arms race' which invited the USA and the Soviet Union to 'adopt a mutual freeze on the testing, production and deployment of nuclear weapons and of missiles and new aircraft designed primarily to deliver nuclear weapons.' The proposal did not cover submarines, which could be replaced at the end of their service life provided that existing rather than new missiles were fitted in the boats. However, other delivery systems (missiles, aircraft) would be maintained rather than replaced.

Advocates of a freeze argue that the time to initiate one is right given the broad parity that currently exists between the superpowers. They believe that a freeze would stop the qualitative arms race. All other proposals allow the technological arms race to continue within numerical constraints. By preventing the introduction of the highly destabilizing counter-force weapons now planned by the USA and the Soviet Union, a freeze would help stabilize the arms race. The fact that it would cover *production* as well as *deployment* of nuclear weapons would prevent 'a production but not deployment race', while the ban on production coupled with a ban on testing nuclear warheads would, over the years, result in a decline in the reliability and readiness of existing nuclear armaments. This, it is claimed, should enhance security by making it less likely that either side would risk starting a nuclear war. Similarly, a ban on the flight-testing of missiles would erode confidence in the use of such weapons against counter-force silos.

Supporters of the freeze point out that it is intended to be the start of a period of intensive negotiations leading to arms reductions and disarmament, and not an end in itself. The Kennedy–Hatfield freeze resolution put before the US Congress in March 1982, for example, called for a freeze to be followed by 'annual percentage reductions or equally effective means' (Kennedy and Hatfield, 1982, p. xviii). Where the freeze proposal is vitally important is that it would end the tendency to introduce new weapons as 'bargaining chips' to be placed on the table for future negotiations and it would also signal to non-nuclear-weapon states the clear intent of the nuclear weapons states to call a halt to vertical proliferation. In essence, proponents of the freeze argue that one should 'freeze first, then reduce', thus stopping the arms race now. Those who favour the more traditional approach argue that one should 'reduce first, then freeze'.

The proposal for a freeze has won widespread support in the USA. In May 1983 the House of Representatives voted 278 to 149 in favour of a freeze followed by mutual, major and verifiable reductions in nuclear arms. The Senate, however, rejected the proposal in the wake of the Soviet shooting down of a Korean air liner. In Britain the proposal has also had widespread support.

The SDP/Liberal Alliance gives general support for a freeze, while the Labour Party supports it as a first step towards disarmament and a 'no-first-use' agreement. The Conservative Government has, however, opposed it.

Opposition to a freeze has concentrated on problems of verifiability. For example, Moreton (1983, p. 57) argues that 'a nuclear freeze is unworkable at the present time because it is unverifiable'. Senators Kennedy and Hatfield (1982) argue that a freeze would be easy to verify because no new weapons could be added to existing arsenals (p. 152). This does not mean that some aspects of the freeze would not be more difficult than others to verify but overall there would be a high probability of identifying breaches. A second argument is that a freeze would confirm Soviet superiority. This is nonsense as the evidence presented in chapter 2 on the nuclear balance indicated (pp. 27–31). A third argument is that the imposition of a freeze would remove any incentive for the Soviet Union to negotiate substantial arms reductions. In fact these are less likely when both sides are adding to their arsenals and trying to pre-empt developments by the other side while securing their own position by putting yet more 'bargaining chips' on the negotiating table.

The Soviet Union has accepted the idea of a freeze. In May 1982 President Brezhnev proposed an immediate freeze on strategic arms deployment and the start of talks on their limitation. On a number of occasions Soviet spokesmen have indicated that they would favour a freeze. It would be in the Soviet interest, firstly because it would enhance the country's security by stopping the move to first-strike weapons; secondly, it would alleviate the economic cost of the arms race; and thirdly, it would help to lower the risk of nuclear war. For the same reasons it would be sensible for the West to endorse a freeze. Unfortunately, neither the USA nor the British governments agree for the reasons outlined above.

Proposals for Force Reductions

When the START talks collapsed, there were two proposals on the table (see table 2.8, p. 34). The Soviet proposals essentially preserved the structure of SALT II while bringing down the ceilings for strategic delivery vehicles. Both sides would have been able to maintain strategic nuclear forces similar to their existing ones and implement their planned modernization programmes (MX, Trident, B-1 bomber, modernized Soviet ICBMs and SLBMs) subject to some restraints in order to remain within the upper limits put forward by the USSR. The Soviet proposals did not meet the US concerns firstly about the strong emphasis on ICBMs in the Soviet force which the USA regard as first-strike weapons, and secondly, the Soviet lead in missile throw-weight. The US START proposals were designed to change this.

Michael Krepon (1983, p. 242) has observed that 'any agreement that does not require a restructuring of Soviet forces raises questions of equity for the Reagan team.' The US proposals of May 1983 greatly favoured the USA. Under the first phase the USA would have been able to pursue its major strategic programmes including MX and additions to the Minuteman III force while the USSR would have had to restructure its forces by cutting ICBMs. The phase II proposals would have posed even more problems for the Soviet Union given the greater throw-weight of their forces. As Krepon (1983, p. 242) commented, 'serious problems of equity are inherent in the negotiations: the Soviet Union sees no equity in an agreement that requires a restructuring of its forces with no apparent impact on US strategic modernization programs.' Not surprisingly, Soviet officials reject totally the idea that the USA can dictate Soviet force structure.

Apologists for the US START proposals acknowledged that they

> would force the Soviet Union to restructure its strategic forces away from the traditional Soviet reliance on ICBM... but is it bad for the Soviet Union?... the vulnerability of land based missiles is a matter over which both sides should be concerned... A primary attribute of the United States proposal is that it will force a relative shift away from ICBMs, resulting in a less destabilising force structure on both sides. This is in the Soviet interest as well as our own (R. Burt, 1982).

Various proposals for 'deep cuts' in current strategic force levels have been made. For example, George Kennan (1981) proposed that there should be an immediate 50 per cent reduction 'across the board... affecting in equal measure all forms of the weapons, strategic, medium range, and tactical, as well as their means of delivery'. Obviously such a proposal would have an immediate effect and considerable public impact. However, given assymetries in US and Soviet force structure, questions of equity would be raised. The Soviet Union would have needed to cut far more launchers than the USA, while the USA would have had to cut more warheads on launchers. Both ICBM forces would be more vulnerable to a pre-emptive first-strike by the other side. Opponents of deep cuts argue that low force levels play into the hands of potential aggressors and make the strategic balance particularly vulnerable to developments in air defence, ASW or BMD developments and a depressed trajectory launch capability which would allow quicker strikes than could be achieved by missiles flying in ballistic trajectories. All such developments, added to a force structure based on first-strike weapons, would make a first-strike more attractive and thus make the strategic balance less stable.

Other suggestions have included proposals that would result in a reduction in the throw-weight of missiles. Such proposals would, however, disadvantage the Soviet Union since Soviet missiles have a much greater throw-weight than their US counterparts. For example, the throw-weight of the SS-18 is 16,500 pounds, that of the SS-19 is 7,000 pounds and that of the SS-17, 6,000 pounds. In comparison the throw-weights of the Minuteman II and III are 1,625 and 1,975 pounds respectively. On the US side only the ageing Titan II missiles with a throw-weight of 8,275 pounds approach Soviet levels and they are being withdrawn.

Another concept that has received some discussion is that of 'build-down'. This was promoted by members of the US Congress (notably Senators Cohen, Nunn and Perry and Representatives Aspin, Gore and Dicks) during 1983 (Report, 1983, pp. 44–5). The basic principle of the proposals is for the nuclear-weapon states to eliminate an agreed number of nuclear warheads from their operational inventories for each new warhead they deploy. The overall purpose of the proposal is to forestall an open-ended arms race while allowing modernization and replacement of ageing and more vulnerable systems. Various 'reductions' formula have been proposed. One of these called for the retirement of two old warheads or bombs for each new ICBM warhead, three old warheads or bombs for two new SLBM warheads and one for each new bomber weapon. This particular formula would lead to proportionately greater reductions in ICBMs and missiles with multiple warheads. In general, the Soviet Union would have to make more significant reductions under such a proposal. A wide range of possible force structures could result from build-down proposals. One criticism of the concept is that unless there are some constraints on the end point, there can be no assurance that the resulting nuclear balance or force postures would be acceptable. Critics also argue that build-down permits the very process which arms negotiations should bring to a halt – the technological modernization of weapons. Build-down proposals were introduced into the START negotiations by President Reagan (October 1983). With the collapse of START these discussions got nowhere and there is no evidence that the USSR is interested in the idea.

One of the problems with force-reduction proposals is that they are put forward within the context of the doctrine of deterrence. Since deterrence rests upon the power of each side to destroy the other after suffering a first-strike, it follows that there is a limit beyond which force reductions cannot go. It can even be argued, as Jonathan Schell (1984, p. 81) observes, that 'overkill' is useful in producing a deterrent effect since 'it may eliminate every last shred of doubt on either side that to make war in a nuclear world is to commit suicide.'

Unilateral or Multilateral?

Unilateral disarmament is stigmatized by government spokesmen in Britain as 'one-sided disarmament' and is often characterized as leaving the disarmed side defenceless in the face of a wholly armed adversary. A leaflet produced by the British Central Office of Information (1981, p. 3) argued that:

> Unilateral disarmament by Britain and its allies is clearly not a safe or sensible alternative. Pressure for unilateral moves will encourage the Russians to block any negotiations in the belief that, if they wait long enough, the West will disarm on its own, without seeking Soviet reductions in return.
>
> Any one-sided reduction by the West would weaken its ability to deter aggression – and therefore make war more likely, rather than less...
>
> The Russians have made it clear that they will not disarm unilaterally.

Those in favour of unilateralism are not necessarily pressing for an immediate renunciation of all nuclear weapons. Many of them are seeking to reverse the arms race by ensuring that unilateral steps towards disarmament are reciprocated, thus leading to further steps. The theoretical basis for this approach was spelt out by Charles Osgood (1962). He focused on a process which he called Graduated Reciprocation In Tension-Reduction (GRIT). This is based on the idea that relatively small-scale unilateral initiatives will help to build-up confidence in the process of disarmament and lead to reciprocation. The nature of particular steps and the speed at which they would be taken depend on the degree of reciprocation. However, the process, once started, must be maintained. It is quite possible that an adversary will at times take advantage of these steps to improve its own military position and in this case remedial action may need to be taken to show that one is not 'soft'. Such remedial action should, however, be directly related to the area in which advantage is being sought and must not itself raise tension. Osgood described the overall aims of GRIT as being 'to reduce and control international tension levels... to create gradually an atmosphere of mutual trust within which negotiations on critical political and military issues will have a better chance of succeeding... to enable this country [the USA] to take the initiative in foreign policy... to launch a new kind of international behaviour that is appropriate to the nuclear age' (Osgood, 1962, p. 88).

'No First-Use'

NATO's policy of 'flexible response' is based on the assumption that the alliance will initiate nuclear war if conventional defence fails to hold a Soviet advance into western Europe.

Those who argue in favour of a no-first-use policy believe that first-use of nuclear weapons will be followed by a nuclear war. As McGeorge Bundy, George Kennan, Robert McNamara and Gerard Smith (1982, p. 32) argued, 'It is time to recognize that no one has ever succeeded in advancing any persuasive reason to believe that any use of nuclear weapons, even on the smallest scale, could reliably be expected to remain limited'. A posture of no-first-use accompanied by increased expenditure on conventional defence would, they argue, firstly reduce the risk of nuclear war and secondly provide an escape 'from many of the complex arguments that have led to assertions that all sorts of new nuclear capabilities are necessary to create or restore a capability for something called "escalation dominance" – and thus lead to the realization 'that the only nuclear need of the Alliance is for adequately survivable and varied *second strike* forces', so that 'requirements for the modernization of major nuclear systems will become more modest' (pp. 37–8).

These arguments were rejected by four West Germans – Karl Kaiser, George Leber, Alois Mertes and Franz-Josef Schultz (1982). They argued that it was precisely fear of nuclear war that had led to the longest period of peace in European history (p. 44). While admitting that this fear has also 'stimulated the build-up of arsenals, since neither side wanted to lapse into a position of inferiority' (p. 44), they argued that 'The coupling of conventional and nuclear weapons has rendered war between East and West unwageable and unwinnable up to now. It is the inescapable paradox of this strategy of war prevention that the will to conduct nuclear war must be demonstrated in order to prevent war at all' (p. 45). A no-first-use policy would effectively limit the existing US nuclear guarantee, greatly increase the significance of Soviet conventional superiority and make it more likely that it would be used, and weaken that NATO Alliance by destroying European confidence in the USA. They did, however, accept that efforts to raise the nuclear threshold by strengthening conventional defence were urgently necessary.

The idea of strengthening NATO's conventional forces has been widely advocated, not least by General Bernard Rogers, the Supreme Allied Commander Europe. The main objection to any strengthening of conventional forces is the cost. However, it is accepted that increased expenditure on conventional force levels would raise the nuclear threshold and as such be consistent with the so-called 'no-early-first-use' stance which has been welcomed by General Rogers.

The debate on a policy of no-first-use has been given added importance by the fact that during the UN Second Special Session on Disarmament the Soviet Union assumed 'an obligation not to be the first to use nuclear weapons' (18 June 1982). Critics argue cynically that a declaratory policy of no-first-use is worthless since it can easily be broken, and that any attempts to introduce certainty into the question of the use or non-use of nuclear weapons weakens deterrence. It is often suggested that a policy of no-first-use, if it is to be credible, must be accompanied by changes in force structure linked to the idea of nuclear disengagement zones. In Central Europe such zones would raise the nuclear threshold by removing battlefield and short-range theatre nuclear weapons (i.e. those with a range of under 200 km) yet maintain NATO's forward defence posture. The Palme Commission, for example, proposed a 300-km battlefield nuclear-weapon zone in Central Europe (Palme Commission, 1982, pp. 147–8) while Lodgaard and Berg (1983, p. 104) have argued that a 200-km zone would cover most of the weapons covered by a 300-km zone (the exception being the Soviet Scud B), be politically easier to implement, and diminish the immediate costs arising from the relocation of military units.

The existing situation in which large stocks of nuclear arms are deployed by NATO close to the East–West frontier and are stored in close proximity with conventional munitions 'is not compatible with the caution and prudence the NATO's decision-making councils are supposed to exercise at the outbreak of hostilities and in the early phases of a war' (Blackaby et al. 1984, p. 19). Atomic demolition mines, for example, would have to be used at the very start of a conflict to avoid being overrun or lost. Indeed, the current disposition of NATO battlefield weapons makes it likely that they will be used earlier rather than later. It is for this reason that a policy of no-first-use or no-early-first-use is usually linked to proposals for a nuclear disengagement zone.

Nuclear disengagement would, however, place greater emphasis on the conventional balance in a situation in which the general (official) view in the West is that the Warsaw Pact has marked superiority. This means that nuclear disengagement probably needs to be linked to either increases in NATO conventional forces (discussed above) or progress in the Mutual Force Reduction Talks. However, Galtung (1984, p. 155) argues that a battlefield nuclear-weapons-free zone would only stimulate the development of conventional offensive systems and longer-range nuclear systems unless it were accompanied by a move towards defensive rather than offensive weaponry. There are also acknowledged problems of verification, particularly in respect of dual capable systems. The imposition of a nuclear disengagement zone would have to be accompanied by changes in force configurations and in military doctrine. It would reduce the danger of inadvertent escalation to nuclear war, and solve in part the 'use them or lose them' dilemma currently facing NATO planners.

Nuclear-Weapon-Free Zones (NWFZs)

The concept of a nuclear-weapon-free zone (NWFZ) rests on three principles:

1 The country or countries forming a NWFZ incur a legal obligation under international law not to produce or deploy nuclear weapons in the territory and not to allow other countries to deploy such weapons. This obligation holds in both peace and war.
2 Nuclear-weapon states for their part undertake to respect the status of the NWFZ and refrain from the use or threat of use of such weapons against the states of such a zone. An implication of this is that nuclear weapons deployed in areas adjacent to the NWFZ and which, because of their range, could only be aimed at targets within the NFWZ, should be removed.
3 A means of verifying that both the member states of a NWFZ and the weapons states comply with their commitments.

There are certain difficulties here. For example, the innocent passage of nuclear weapons through the territorial seas or airspace above such seas would be difficult, if not impossible, to stop. Problems are also raised by courtesy visits by nuclear armed vessels to ports in the NWFZ and by cruise missiles which would violate the airspace of a NWFZ if fired across its territory. Ballistic missiles, on the other hand, are not held to violate airspace.

Existing nuclear-weapon-free zones exist in Antarctica under the Antarctic Treaty (1959) and in Latin America (Treaty of Tlatelolco, 1967). Unfortunately, while the nuclear-weapon states, under Additional Protocol II to the Treaty of Tlatelolco, have undertaken to respect the military denuclearized status of Latin America and not to contribute to acts involving a violation of the Treaty, nor to use or threaten to use nuclear weapons against the parties of the Treaty, both the USA and Britain have reserved the right to reconsider their obligation with regard to a state in the NWFZ in the event of any act of aggression or armed attack by that state carried out with the support or assistance of a nuclear-weapon state.

Various proposals for further NWFZs have been put forward, for example in Africa, the Middle East, South Asia and the South Pacific, all of which have been the subject of resolutions in the UN General Assembly. Attention has also been focused on plans for a nuclear-weapon-free zone in the South Pacific where leaders of 14 states unanimously agreed in August 1984 that their region should become nuclear free and further agreed to establish a working party to draft a nuclear-free-zone treaty and report back to the leaders of the 'Pacific Forum' in 1985 (*The Guardian*, 28 August 1984). The Government of Fiji has, however, put these proposals in doubt by promising that nuclear-armed US warships will be welcome at any time in the dockyards of Suva (*The Sunday Times*, 30 December 1984).

Attention has also focused on various proposals for NWFZs in Europe. The origins of such proposals go back to the early 1950s and the search for a political solution to the German question. In the early 1950s the only alternative to a nuclear build-up in central Europe was believed to be a reunification of Germany on the basis of neutrality. Soviet proposals to this effect were rejected by the West which feared that a reunited Germany might come under strong Soviet influence. The Warsaw Pact then proposed the establishment of a denuclearized zone comprising the two German states (January 1956). Various versions of the plan were put forward between then and the end of 1958, all of which were rejected by the West.

These ideas sparked proposals for further NWFZs on the flanks of Europe. In January 1958 Soviet Prime Minister Bulganin proposed to Norway and Denmark (the two Scandinavian members of NATO) that an NWFZ should be established in Northern Europe. The idea was taken up by Khruschev in a speech delivered on 11 June 1959 in which he proposed the inclusion of the Baltic in the zone. Under these proposals nuclear weapons were not to be stationed in the area, and in 1961 Norway and Denmark responded to these initiatives by declaring that they would not allow the stationing of nuclear weapons on their territory in time of peace. Subsequent plans were proposed by the Swedish government (the Undén Plan) and by President Kekkonen of Finland without success.

Since 1978 there has been increased interest in the proposals for a Nordic NWFZ. Generally speaking, current proposals do not include Soviet territory (notably the Kola Peninsula and the Baltic) although the Swedish government has some reservations on this score. Some proposals envisage extension of the Nordic NWFZ to cover all the areas controlled by the Nordic States including Greenland, Iceland, Spitsbergen, Jan Mayen Island, the Faeroe Islands, Aaland and Bear Island. Both Spitsbergen (Norwegian) and Aaland (Finnish) have been demilitarized by treaty for several decades. Iceland is a member of NATO but has no military forces of its own. Under a 1951 treaty the defence of the country is the responsibility of the USA. The defence of Greenland is a joint Danish–American responsibility. The presence of US bases in Greenland and Iceland raises problems as does the position of Norway and Denmark. Their participation in a NWFZ would require them to take an unqualified position against deployment of nuclear weapons in wartime. This would effectively decouple Norway and Denmark from NATO's nuclear strategy. Danish participation in a Nordic NWFZ would also disrupt current NATO plans for the defence of Denmark, Schleswig-Holstein, Hamburg and the Danish straits under the joint West German/Danish Commander Allied Forces Baltic Approaches.

A further difficulty is raised by the question of the maritime extent of the proposed Nordic NWFZ. Territorial waters only extend 12 miles from the coast, although it has been suggested that NWFZs might extend as far as the 200 nautical-mile exclusive economic zone as defined in the Law of the Sea Convention. Proposals for a Nordic NWFZ have to contend with the fact that the North Atlantic, Norwegian Sea and Barents Sea have much greater strategic significance now than was the case twenty years ago (Miller, 1983, p. 117). Such developments have clear implications for any proposals to make Norway part of a NWFZ.

Proposals for a Balkan denuclearized zone were put forward by Romania in 1957 and taken up by Khrushchev, who in May 1959 proposed an NWFZ in the Balkan and Mediterranean areas. It has since been revived with general support from the governmnents of Bulgaria, Greece, Romania and Yugoslavia but has been rejected by Turkey – a move that is likely to end Romanian and Bulgarian interest in the plan (*The Sunday Times*, 3 February 1985).

Finally, during 1980, the Russell Foundation joined forces with a number of organizations in Britain (CND, Pax Christi and others) to launch a European appeal for nuclear disarmament. An early draft of the appeal was circulated to groups in Europe before being redrafted and promulgated on 28 April 1980 following an international meeting held in London. The appeal called for the people of Europe 'to act together to free the entire territory of Europe, from Poland to Portugal, from nuclear weapons, air and submarine bases, and from all institutions engaged in research into or manufacture of nuclear weapons' ('Appeal for European Nuclear Disarmament', in Thompson and Smith, 1980, p. 224). Out of this appeal grew the European Nuclear Disarmament (END) movement which led in turn to increased contacts between European peace movements (Coates, 1980). The END appeal was dismissed by Lawrence Freedman (1980, p. 3) as 'frivolous and pernicious': 'The strategy of END appears to be to foment sufficient protest and outrage in both East and West to undermine the alliance system... I can think of no single event more likely to stimulate World War III than a successful revolt against the Warsaw Pact.' The danger to world peace, and the slide towards the holocaust, comes not from new weapons or from 'variations on the theme of limited nuclear war' but from 'political conflict and change' (p. 3).

Responding to Professor Freedman, Mary Kaldor argued that the appeal for a European nuclear free zone from Poland to Portugal was deliberately chosen to 'appeal to people within the *political* territory of Europe' who 'do not want to be drawn into a war between the superpowers for which they have no reason and over which they have no control' (Kaldor, 1981, p. 3).

Those in favour of NWFZs and zones of disengagement argue that they would:

1 institutionalize policies not to station nuclear weapons in certain areas;
2 function as a buffer or disengagement zone between rival power blocs in strategically vital areas:
3 operate as pivotal confidence building arrangements, promoting the containment of nuclear arms;
4 serve as an impulse to effective arms control and limitation;
5 stimulate détente;

6 reduce the possibility of war;
7 positively increase the security of the countries or areas incorporated into the NWFZ.

Opponents of NWFZs argue that while it is possible to draw lines around a geographical area and ban the emplacement of weapons there, 'it is not possible to safeguard any such zones from the use of nuclear weapons from outside. Nuclear-free zones can be created at the stroke of a pen: there is no such thing as a nuclear-safe zone' (Moreton, 1983, p. 63).

Non-Nuclear Defence

Unilateral nuclear disarmament is generally recognized as not being a viable policy option for the superpowers. In their case a freeze followed by progress towards mutual disarmament is generally felt to be the most profitable avenue for progress. So far as Europe is concerned, interest has focused on the development of non-nuclear defence policies and on the idea that Europe might adopt a neutralist position *vis-à-vis* the superpowers.

Non-Nuclear Defence Strategies

This is not the place to examine in detail the various non-nuclear defence strategies which have been adopted, notably in Sweden, Switzerland and Yugoslavia, or which might be adopted by other European countries. Broadly speaking there are four possible elements to such a policy.

Frontier defence

There is little call for offensive weaponry if the primary purpose is to maintain the territorial integrity of the country rather than defeat the enemy. The main emphasis then falls on defeating an enemy advance.

Developments in emerging technology (ET) weaponry, including increasingly accurate short-range missiles designed to destroy tanks, aircraft and ships, have reinforced the traditional advantage of defence and suggest that, by investing heavily in precision guided munitions (PGMs), NATO might be able to counter a

Waraw Pact advance while adopting a purely defensive posture. While counter measures can be taken against PGMs, their overall effect seems to be:

1 to challenge the superiority of the tank on the ground (e.g. 1973 Arab–Israeli War);
2 to endanger large surface ships (e.g. Falklands War);
3 to make aircraft vulnerable (air-to-air and surface-to-air missiles) while increasing the accuracy of air-to-ground munitions (e.g. laser guided bombs) once the aircraft get through defences.

Non-nuclear frontier defence would probably be insufficient to act as a deterrent on its own. It would need to be combined with plans for territorial defence. The main purpose of a strong frontier defence would be to demonstrate a clear will to resist invasion and to extract as high an 'entry price' as possible.

Territorial defence

Territorial defence has been defined by Adam Roberts (1976, p. 34) as:

> a system of defence in depth: it is the governmentally-organised defence of a state's own territory. It is aimed at creating a situation in which an invader, even though he may at least for a time gain geographical possession of part or all of the territory is constantly harassed and attacked from all sides. It is a form of defence strategy which has substantial reliance on a citizen army, including local units of a militia type. Characteristically a territorial defence system is based on weapons systems, strategies, and methods of military organisation which are better suited to their defensive role than to engagements in major military actions abroad.

Various proposals for territorial defence have been put forward. A small Study Group on Alternative Security Policies at the Max Planck Institute in Starnberg, West Germany developed a proposal for a new territorial defence structure based on small infantry units dispersed throughout the countryside, on average 3–4 men per square kilometre, who would always be stationed on the same territory, would know it intimately, and would therefore be able to prepare to defend it utilizing anti-personnel weapons, mines and cheap short-range missile launchers and anti-tank weapons. The cardinal principle of the proposal is that these groups would not provide a large target to the enemy, thus encouraging use of Soviet nuclear weapons. The units would be supported by a precision-guided artillery rocket network which they could call on to attack larger enemy concentrations. The ideal system would consist of guided missiles fired from cheap and expendable launchers which could be abandoned once they had been used (since they would be targeted by the enemy's artillery once used). The rockets would be guided on to their target by the dispersed infantry units, utilizing an information network which would link them with other units and with the artillery network and the high command (Afheldt, 1983).

These proposals are based on the idea that the enemy should not be offered any military targets worth attacking with heavy means of destruction (such as aircraft, airports, tanks, heavy infantry vehicles, large missile systems and heavy concentrations of troops). NATO's current forward-defence strategy has a large number of such targets and attacks on these targets using nuclear weapons would bring about the destruction of West Germany.

Protracted guerrilla warfare

Protracted guerrilla warfare relies on a process of attrition to wear down the enemy forces and reduce their effectiveness. In some circumstances it might complement conventional defence, in others it would be an alternative that could continue following a collapse of the regular army. Defence by guerrilla warfare would greatly reduce the likelihood of nuclear weapons being used by an opponent. The fact that a country was known to be prepared for protracted guerrilla warfare would also act as a deterrent.

Various categories of guerrilla warfare are possible including, firstly, partisan warfare, where professional or semi-professional forces operate from rural strongholds in a military campaign which ultimately leads to the defeat of the enemy in conventional warfare (e.g. Vietnam, Cuba); secondly, urban guerrilla warfare involving sabotage of military and economic targets and attacks on military personnel to sap the morale and will of the opponent (e.g. the underground resistance in northern Europe during the Second

World War); and, thirdly, political terrorism, including the indiscriminate use of violence for political ends aimed at, among others, civilians of the occupying power and one's own civilians who collaborate with the enemy.

Partisan warfare is probably not possible in Britain or in much of Western Europe. The most realistic option therefore remains urban guerrilla warfare. Combined with non-violent civil resistance, it could help to make it very difficult to govern a country. The limitations of guerrilla warfare stem from the protracted nature of the struggle, the likelihood of reprisals by the occupying power, the thin line that exists between terrorism and guerrilla warfare and the fact that guerrilla warfare itself is unlikely to lead to military victory against an occupying power. However, in the face of concerted guerrilla warfare across Western Europe in combination with other forms of unrest, withdrawal might become an attractive option to an occupying power.

Non-violent civil resistance

Non-violent civil resistance is resistance by the civilian population in the form of strikes, boycotts, civil disobedience, mass non-cooperation, etc. Such tactics have been used in India (during the struggle for independence), in resistance to the Franco-Belgium occupation of the Ruhr in 1933–5 and the Soviet/Warsaw Pact invasion of Czechoslovakia in 1968, and by the Solidarity movement in Poland. Civil resistance can make government of a country difficult if not impossible and thus deny the occupying power some of the benefits it hoped to derive from the occupation.

The prospect of having to face determined political opposition and mass non-cooperation would increase the general deterrent effect although by itself civil resistance has distinct limitations.

Neutralism

Many of those critical of current defence policy argue that European interests would be better served if Europe could break away from its entangling alliances with the superpowers and adopt a neutralist position. Galtung, for example, argues that 'from the point of view of the superpowers an alliance is tantamount to an extension of superpower territory for military purposes' (1984, p.

185). For Galtung, the threat to European security arises from its coupling with the superpowers. He proposes 'gradual decoupling from the superpowers... as one way of increasing security' (Galtung, 1984, pp. 184–92). In this sense neutralism needs to be distinguished from the idea of nuclear disengagement in Europe under which European states would be 'denuclearized' yet remain within the existing NATO and Warsaw Pact alliances. Neutralism aims ultimately at the dissolution of the post-war alliances.

Given the systemic nature of the conflict between the superpowers, the neutralist option is regarded by its critics as defeatist. It is also argued that West European neutralism would be little more than a prelude to Soviet control of the continent. Former Italian Defence Minister Taviani expressed this fear when he argued that 'there is just one step from denuclearisation to neutrality, and the step from neutralisation to Sovietisation would be even shorter' (quoted in Albrecht, 1982, p. 157). An alternative argument is to suggest that neutralism would lead to 'Finlandization' under which Western Europe as a whole would be unduly influenced by the Soviet Union. Those who dismiss this fear point out that Finland's position derives from the fact that it was for a while part of the Russian Empire; that Finland is far from being dominated by the Soviet Union; and that the collective neutralism of Western Europe would offer a greater freedom of action than membership of the alliances.

Neutralism does not necessarily mean that each European country would be required to stand on its own. It could involve some kind of European-wide defence association independent of the superpowers. Galtung (1984, p. 190), for example, sees 'many positive aspects in maintaining, in the West, a Western defence alliance' based on 'defensive defence' and reflecting a concern for collective security which is not, as is the present NATO policy, 'provocative and unstable' and 'much too easily combined with planning for attack'.

Various alternatives to continuing membership of NATO have been suggested including a non-nuclear European defence association. One argument against this option is that such an association would have even greater problems in mounting a credible defence against the Warsaw Pact if it tried to do this without the USA. Philip Towle (1983) argues that an alliance of the European members of NATO without the USA would be hard pressed to defend Western Europe against Warsaw Pact troops. It would have to defend an arc of countries against a geographically more compact alliance with better internal lines of communication. Some Western European forces (e.g. those of Spain and Portgual) are ill-positioned to meet an initial Soviet thrust. The Western European countries would also suffer because their equipment is often not compatible, while that of the Warsaw Pact is. The Soviet Union has marginally more troops than the NATO countries combined. While it falls very far short of the 3 : 1 advantage usually thought to be required by attacking forces, it could pose a significant threat to Western European air forces and to the sea lanes supplying Europe, and has, of course, a significant lead in armour. The Western European lead in anti-tank weapons is not, however, mentioned. Towle argues that the Western Europeans would need to spend much more if they were to defend themselves without US help and even then they 'would not be able to deter a Soviet attack as effectively as NATO can today' (Towle, 1983, p. 28).

Another proposal is for the formation of a fairly loose association that would rely primarily on the national defence of each member state, but guarantee certain kinds of assistance in the event of an attack. Within the Alternative Defence Commission such ideas were greeted with scepticism, although it was agreed that 'if the attempt to de-nuclearise NATO was obstructed by the USA, and there was strong European support for a non-nuclear defence organisation, this could clearly not be ruled out as an option' (Alternative Defence Commission, 1983, p. 100). Such a European defence association could, moreover, stress its defensive nature by emphasizing territorial defence, guerrilla warfare and civil resistance, either separately or in some combination.

A Policy for Britain

Those who advocate a non-nuclear defence policy for Britain start from the position that Britain is a certain target for a Soviet nuclear strike because it is both a nuclear power in its own right and a base for American nuclear forces. The effects of a nuclear attack on a small heavily populated island would, it is said, be devastating. It is said that Britain's security interests would be better served if Britain adopted a non-nuclear defence policy.

Firstly, as a neutralist non-nuclear state Britain's significance in future superpower conflicts would be greatly diminished. The chances of it being attacked would be low. Secondly, there would in any case be no point in attacking a non-nuclear Britain with nuclear weapons since this would destroy the very facilities the enemy would wish to use. Thirdly, the Soviet Union has said that it will not attack a non-nuclear country with nuclear weapons. Whether or not it would keep this promise is not clear but the uncertainty here is better than the certainty of knowing that Britain could be destroyed by nuclear weapons because it harbours such weapons itself.

Various assessments have been made of the size and structure of Britain's military forces within the context of a non-nuclear defence policy. The Alternative Defence Commission (1983, pp. 282–4) touched on this issue, and proposals have also been made by Smith (1982, pp. 13–20), Smith and Gapes (1984, pp. 21–9) and Johnson (1985, pp. 91–110). The starting point for such assessments is an analysis of the objectives Britain's defence policy should seek to achieve. There seems to be general agreement among those advocating alternative defence policies that these should include the following.

1 The territorial defence of the United Kingdom, although even here the Alternative Defence Commission pointed to 'grey areas' – notably Northern Ireland.
2 The protection of British resources in the surrounding waters, as defined by international agreement, including oil, gas, mineral and fishing rights.
3 The maintenance of the security and freedom of maritime trade routes between the United Kingdom and the rest of the world and, in particular, opposition to any attempt to impose a blockade on the United Kingdom.
4 Countering internal threats to democratic self-government, if necessary by the internal use of military force in defence of the values and human rights that underpin such societies. There are nevertheless problems since there is a thin line between the defence and the subversion of democracy.

To achieve the defence policy objectives outlined above, Britain should, it is suggested, concentrate on frontier and territorial defence coupled with increased emphasis on civil defence and some consideration of civilian defence forces. As an island, Britain enjoys certain advantages in respect of frontier defence which are not shared by its European allies. Frontier defence for Britain would require air defence (fighters and surface-to-air missiles), sea defences (particularly fast patrol craft, land-based aircraft and submarines), and mobile ground forces suitable for counter-attacks against any forces that landed on British soil. Rejection of an overseas defence role would enable Britain to dispense with long-range tanker aircraft and large surface vessels. The kind of force structures envisaged by those pressing for such a change in British defence policy include the following.

1 For the army (a) reduced emphasis on heavy armour, (b) increased emphasis on regular defence by infantry using precision guided munitions and supported by helicopters to provide air mobility and (c) increased emphasis on in-depth defence based on territorial units to extract a high entry price for invasion.
2 For the navy, a sharp reduction in surface vessels, coupled with the retention of non-nuclear armed submarines and a marked increase in patrol craft, minelayers and minesweepers, thus emphasizing the navy's role in frontier defence.
3 The contribution of the RAF to air support would be primarily defensive. There would be a marked reduction in strike/attack squadrons. Ground support and interceptor squadrons would be retained as would an airborne early warning capability, reconnaissance and maritime patrol capability. Also retained would be maritime support aircraft including those providing air cover and a maritime strike capability.
4 As well as an increased role for territorial defence, consideration could be given to the development of a civilian defence force, perhaps based on the Swiss model, and to preparation for non-violent civilian resistance in the event of an invasion.
5 Increased emphasis on civil defence which, within the context of a non-nuclear defence policy, would make sense.

Britain's contribution to European defence raises considerable problems. Participation in the NATO alliance is a major cornerstone of current policy and is achieved through the maintenance of ground and tactical air forces in West Germany, participation in NATO planning and military exercises, and through a large and continuing naval contribution in the North Atlantic. Many of those who have recently questioned Britain's defence policies have, nevertheless, concluded that Britain should remain within NATO while seeking to change NATO's defence policy from a nuclear to a non-nuclear one. The Church of England's report *The Church and the Bomb* (CEBSR, 1982) argued that British unilateral nuclear disarmament within the context of continued membership of NATO could be seen as 'a unilateral stage within a multilateral process' (p. 139) whereas renunciation coupled with withdrawal from NATO would result in a relatively small reduction in nuclear weapons and might be seen by the Soviet Union as 'the possible beginning of a break-up of NATO' and thus fail to have 'the creative effect on negotiations of a unilateral gesture within the framework of NATO itself' (p. 139).

The Alternative Defence Commission (1983) has argued that British security could be imperilled if Western Europe were occupied or controlled by a foreign power and that Britain has an interest in the collective defence of Western Europe. The commission felt that there were 'a number of positive grounds for believing that a Britain without nuclear weapons could promote disarmament and disengagement more effectively by staying in the Alliance' (p. 88) but that given its fundamental rejection of nuclear weapons strategy (including the British deterrent, forward-based US theatre weapons and reliance on the US strategic nuclear umbrella) 'the prima facie case for withdrawing from NATO is compelling' (p. 89). The Commission therefore argued that a policy of staying in NATO should be contingent, firstly, on adoption by NATO of a policy of no first-use of nuclear weapons; secondly, on a phased withdrawal of all battlefield nuclear weapons; thirdly, on removal of all US theatre nuclear weapons; and, fourthly, on decoupling NATO strategy from the US nuclear deterrent by declaring publicly that the European members of the alliance did not wish the USA 'in any circumstances to use nuclear weapons in response to Soviet *conventional* attack even in the event of a Soviet victory' (pp. 92–3). Most of those proposing the denuclearization of NATO argue that a timetable for achieving the goal of a nuclear-free NATO must be set, and that Britain's continued membership of NATO must be contingent on the agreed schedule for denuclearization being met.

The case against membership of NATO rests on NATO's nuclear-orientated defence policy and the fact that NATO aligns Europe to the USA in the rivalry between the superpowers and elsewhere. John Cox, a Vice-President of the Campaign for Nuclear Disarmament (CND), argued that 'withdrawal from NATO is not an end in itself'. Continuing British membership rests largely on whether or not NATO 'proves to be the major obstacle to nuclear disarmament' (Cox, 1982, p. 35).

Peter Johnson, a retired Air Force Squadron Leader and past President of the local Conservative Association in Dartford, Kent, has argued that Britain is now a second rank power which, 'whether we like it or not ... in itself ... is neither a target nor a factor of any real consequence in the calculations of the greatest powers' (Johnson, 1984, p. 79). He argues that 'Risks consist of the danger of being involved in a war and, secondly, of the degree of disaster which might follow such involvement. Protection lies in the possibility that the orientation [of foreign policy] will avoid war altogether' (p. 79). Alliances 'are nearly always the product of a foreseen possibility of war' (p. 79) and he suggests that for Britain 'the real total of risk in a stance of neutrality turns out to be less than that in alliance, and the degree of protection to be greater' (p. 82). In view of this Johnson suggests that 'a change in foreign-policy orientation from alliance to military non-alignment would be the best way' to achieve 'the safety and integrity of our home territory and people' (pp. 83–4). This leads him to propose British withdrawal from NATO. So far as the independent British deterrent is concerned, he argues that none of the arguments for its retention are particularly compelling (p. 95), while the USA would probably not continue to support the British submarine-based system if Britain were neutral. The procurement of a British system would be very expensive. He leaves open the possible retention of tactical nuclear weapons although he suggests that 'on an objective view of the balance of risk/protection involved' they are not likely to be retained (p. 107).

Conclusions

'General and Complete Disarmament' is the technical term usually taken to mean a multilateral reduction in arms throughout the world to levels adequate only for countries' internal policing and maintenance of internal order, together with contributions to a UN force, all carried out under strict and effective international control. Since the late 1950s the prospects of this coming about have, however, seemed remote. The 1960s and 1970s saw major efforts put into *arms control*, not disarmament. Arms control does not necessarily imply *any* reduction in forces. It aims, firstly, to constrain the deployment of military power (although conceivably allow more than is currently deployed) (e.g. SALT); secondly, to limit certain aspects of military activity by banning certain weapons (e.g. biological ones) or by limiting the geographical area in which certain arms are deployed (e.g. Antarctica); and, thirdly, by restraining certain activities in some way (e.g. nuclear testing).

Overall the record of arms control during the 1960s and 1970s must be judged to have been poor. It did not halt the arms race and during President Reagan's first term of office there was little evidence that the USA was seriously interested in pursuing it.

So far as Europe is concerned, a major debate on defence policy is now underway. Two issues are at the centre of this debate: firstly, the validity of the current nuclear-based defence policies and the viability of the various non-nuclear options; and, secondly, the desirability of breaking away from the superpowers and adopting a neutralist position. A range of important and complex issues are involved in these two issues and the debate is by no means over. It is also far from clear that the European nations will be able to evolve a common defence policy. What is important is that the various options should be explored thoroughly and that Britain should develop a policy that meets its own needs.

References and Further Reading

Afheldt, H. (1983) *Defensive Verteidigung*, Reinbek-Hamburg: Rowohlt.

Albrecht, U. (1982) 'Western European Neutralism', in M. Kaldor and D. Smith (eds.) (1982) *Disarming Europe*, London: The Merlin Press.

Alternative Defence Commission (1983) *Defence without the Bomb. The Report of the Alternative Defence Commission set up by the Lansbury House Trust Fund*, London: Taylor and Francis.

'Appeal for European Nuclear Disarmament', in E.P. Thompson and D. Smith (eds) (1980) *Protest and Survive*, Harmondsworth: Penguin.

Arbatov, G.A. (1981) 'The Strategy of Nuclear Madness', *Co-Existence*, 18 (2), 162–74.

Aspin, L. (1979) 'The Verification of the SALT II Agreement', *Scientific American*, 240 (2), 30–7.

Blackaby, F., Goldblat, J. and Lodgaard, S. (eds) (1984) *No-First-Use*, London: Taylor and Francis.

Blackaby, F., Goldblat, J. and Lodgaard, S. (1984) 'No-first-use of nuclear weapons – an overview', in Blackaby, F., Goldblat, J. and Lodgaard, S. (1984) *No-First-Use*, London: Taylor and Francis.

Brown, H. and Davis, L.E. (1984) 'Nuclear Arms Control: Where do we stand?', *Foreign Affairs*, 62 (5), 1145–60.

Bundy, McG., Kennan, G.F., McNamara, R.S. and Smith, G. (1982) 'Nuclear Weapons and the Atlantic Alliance', *Foreign Affairs*, Spring 1982, republished in Blackaby, F., Goldblat, J. and Lodgaard, S. (1984) *No-First-Use*, London: Taylor and Francis.

Burt, R. (1982) 'The Evolution of the United States Start Approach', *NATO Review*, 4.

Central Office of Information (1981)) *The Balanced View. Nuclear Weapons and Arms Control*, London: Arms Control and Disarmament Research Unit, Foreign and Commonwealth Office.

CEBSR (Church of England, Board for Social Responsibility) (1982) *The Church and the Bomb. Nuclear Weapons and Christian Conscience*, London: Hodder and Stoughton.

Coates, K. (1980) 'For a Nuclear-free Europe', in Thompson, E.P. and D. Smith (eds) (1980) *Protest and Survive*, Harmondsworth: Penguin.

Cox, J. (1982) *No No NATO*, London: CND.

Dahlitz, J. (1983) *Nuclear Arms Control*, London: George Allen and Unwin.

Durie, S. and Edwards, R. (1982) *Fuelling the Nuclear Arms Race*, London: Pluto Press Ltd.

Epstein, W. (1976) *The Last Chance. Nuclear Proliferation and Arms Control*, New York: The Free Press.

Forsberg, R. (1982) 'A Bilateral Nuclear-Weapon Freeze', *Scientific American*, 247 (5), 32–41.

Freedman, L. (1980) 'A Criticism of the European Disarmament Movement', *ADIU Report*, 2 (4), 1–4.

Frei, D. (1983) *Risks of Unintentional Nuclear War*, London: Croom Helm.

Galtung, J. (1984) *There are Alternatives! Four Roads to Peace and Security*, Nottingham: Spokesman.

Gilpatric Committee (1965) *Committee on Nuclear Proliferation, Report to the President*, Gilpatric Papers, John F. Kennedy Library, Boston, Mass., USA.

Goldblat, J. (1984) 'US and Soviet allegations of breaches of arms control agreements', in SIPRI (1984) *The Arms Race and Arms Control 1984*, London: Taylor and Francis.

IAEA (1977) Special Safeguards Implementation Report, International Atomic Energy Authority, document COV/1842.8, dated June 1977.

International Treaties:

Multilateral

The Antarctic Treaty, 1959.

Treaty Banning Nuclear Weapon Tests in the Atmosphere, in Outer Space and Under the Water, 1963 (Partial Test Ban Treaty).

Treaty on Principles Governing the Activities of States in the Exploration and Use of Outer Space, including the Moon and other Celestial Bodies, 1967 (The Outer Space Treaty)

Treaty for the Prohibition of Nuclear Weapons in Latin America, 1967 (Treaty of Tlatelolco).

Security Council Resolution on Security Assurances to Non-nuclear Weapon States, 1968. UN Document S/RES/225, 19 June.

Treaty on the Non-Proliferation of Nuclear Weapons, 1968 (Non-Proliferation Treaty).

Treaty on the Prohibition of the Employment of Nuclear Weapons and Other Weapons of Mass Destruction on the Seabed and the Ocean Floor and in the Subsoil Thereof, 1971 (The Seabed Treaty).

Bilateral

US–Soviet Memorandum of understanding regarding the establishment of a direct communications link, 1963 (US–Soviet Hot Line Agreement).

British–Soviet Agreement on the establishment of a direct communications link, 1971 (British–Soviet Hot Line Agreement).

Agreement on measures to improve the USA–USSR direct communications link, 1971 (amended 1975) (US–Soviet Hot Line Modernization Agreement).

Agreement on measures to reduce the risk of outbreak of nuclear war between the USA and the USSR, 1971 (US–Soviet Nuclear Accidents Agreement).

US–Soviet Treaty on the limitation of anti-ballistic missile systems, 1972 (SALT ABM Treaty).

US–Soviet Interim Agreement on certain measures with regard to the limitation of strategic offensive arms, 1972 (SALT I Agreement).

Protocol to the US–Soviet Treaty on the limitation of anti-ballistic missile systems, 1974.

US–Soviet Treaty on the limitation of underground nuclear weapon tests, 1974 (Threshold Test Ban Treaty).

Joint US–Soviet Statement on the question of further limitations of strategic offensive arms, 1974 (Vladisvostok Accord).

US–Soviet Treaty on underground nuclear explosions for peaceful purposes, 1976 (Peaceful Nuclear Explosions Treaty).

French–Soviet Agreement on the prevention of the accidental or unauthorized use of nuclear weapons, 1976 (French–Soviet Nuclear Accidents Agreement).

British–Soviet Agreement on the prevention of an accidental outbreak of nuclear war, 1977 (British–Soviet Nuclear Accidents Agreement).

Treaty between the USA and the USSR on the limitation of strategic arms, 1979 (SALT II Treaty).

Protocol to the SALT II Treaty, 1979.

Johnson, P. (1984) *Neutrality: A Policy For Britain*, London: Maurice Temple Smith.

Kaiser, K., Leber, G., Mertes, A. and Schulze, F-J. (1982) 'Nuclear Weapons and the Preservation of Peace', *Foreign Affairs*, Summer 1982, republished in Blackaby, F., Goldblat, J. and Lodgaard, S. (1984) *No-First-Use*, London: Taylor and Francis.

Kaldor, M. (1981) 'Why we need European Nuclear Disarmament', *ADIU Report*, 3 (1), 1–4.

Kennan, G. (1981) cited in the *Washington Post*, 24 May.

Kennedy, E. and Hatfield, M. (1982) *Freeze! How you can help prevent nuclear war*, New York: Bantam Books.

Krepon, M. (1983) 'Assessing Strategic Arms Reduction Proposals', *World Politics*, 35 (2), 216–44.

Leitenberg, M. (1978) 'Arms Control and Disarmament. A Short Review of a Thirty Year Story', Ottawa: Carleton University. Mimeo.

Lodgaard, S. and Berg, P. (1983) 'Disengagement and nuclear weapon-free zones: raising the nuclear threshold', in S. Lodgaard and M. Thee (eds) (1983) *Nuclear Disengagement in Europe*, London: Taylor and Francis.

Miller, S. (1983) 'The northern seas in Soviet and US strategy', in S. Lodgaard and M. Thee (eds) (1983) *Nuclear Disengagement in Europe*, London: Taylor and Francis.

Moreton, E. (1983) 'Untying the Nuclear Knot', in G. Segal et al. (1983) *Nuclear War and Nuclear Peace*, London: The Macmillan Press.

Myrdal, A. (1980) *The Game of Disarmament. How the United States and Russia run the Arms Race*, Nottingham: Spokesman.

Noel-Baker, P. (1958) *The Arms Race*, London: John Calder.

Osgood, C. (1962) *An Alternative to War and Surrender*, Urbana: University of Illinois Press.

Palme Commission (1982) *Common Security: A Programme for Disarmament*, London: Pan Books.

Pringle, P. and Spigelman, J. (1982) *The Nuclear Barons*, London: Sphere Books Ltd.

Report of the President's Commission on Strategic Forces, April 1983.

Roberts, A. (1976) *Nations in Arms: The Theory and Practice of Territorial Defence*, London: Chatto and Windus.

Schell, J. (1984) *The Abolition*, London: Jonathan Cape.
Senate Committee on Foreign Relations (1979) *The SALT II Treaty. Hearings, July 16–19, 1979*, Washington DC: US Government Printing Office.
Smith, D. (1980) *The Defence of the Realm in the 1980s*, London: Croom Helm.
Smith, D. (1982) *Non-Nuclear Military Options for Britain*, Bradford: University of Bradford School of Peace Studies, London: Housmans.
Smith, P. and Gapes, M. (1984) *Britain without the Bomb*, Leeds: Independent Labour Publications.
SIPRI (Stockholm International Peace Research Institute) (1978) *Arms Control: A Survey and Appraisal of Multilateral Agreements*, London: Taylor and Francis.
SIPRI (1982) *The Arms Race and Arms Control*, London: Taylor and Francis.
SIPRI (1983) *The Arms Race and Arms Control 1983*, London: Taylor and Francis.
Towle, P. (1983) *Europe without America: Could we defend ourselves?*, London: Institute for European Defence and Strategic Studies.

11
Conclusions

The controversy over the neutron bomb, the deployment of cruise and Pershing II missiles, the British decision to acquire Trident, the arms build-up initiated by the Carter Administration and continued under President Reagan, and the failure of the various arms-control talks, has led to a major resurgence of anti-nuclear peace groups in the West. Membership of the British Campaign for Nuclear Disarmament rose from just over 3,200 in 1978 to over 100,000 national members in 1984 with perhaps 140,000 further members belonging to local CND-affiliated organizations. The various groups are by no means united on policy. Some have very wide objectives, others have fairly narrow aims. Some are pacifist, others urge adoption of a non-nuclear defence policy. Some are opposed to the NATO alliance, others would prefer to see continuing British membership of it. All are critical of government policy. To this extent they differ markedly from the official peace groups in the Soviet Union and Eastern Europe, which are recognized by their governments and act as organs for the dissemination of official policy. The few autonomous groups in the East are both small and subject to harassment.

Western governments have responded to the peace movement in a number of ways. An early response was to dismiss them as naive, emotive and irrational, if well-meaning, critics of government policy. As the movements grew in size this became more difficult to sustain, although echoes of this approach could be seen in a speech delivered by British Minister for the Armed Forces Peter Blaker (1983): 'Defence of the people will be decided by cool judgement, not by the shouting of empty slogans on the streets or the carnival cavortings of woolly people in woolly hats'.

The most significant feature of the current peace movement has been the emergence of specialist and professional groups (for example, Generals for Peace and Disarmament, Scientists Against Nuclear Arms, Lawyers versus the Bomb). This has made it increasingly difficult to dismiss the peace movement as naive and ill-informed. Government response has been to change tactics. It has become commonplace to charge that the peace movement is knowingly or unknowingly controlled in some way by the Soviet Union and serves Soviet purposes. Thus John Nott, when Secretary of State for Defence, argued that 'If the Soviet Union feel they can achieve all their objectives through the peace movements in the West [in this case non-deployment of cruise and Pershing IIs], there is no need for them to come forward with proposals to lower their own volume of nuclear weapons' (Nott, 1982). A study of peace movements in Britain and West Germany prepared by the Conservative Bow Group and the Konrad Adenauer Stiftung (1983) alleged that CND's leadership is 'mainly composed of Communist Party members and members of the Tribune Group'. A pamphlet prepared by the Campaign for Defence and Multilateral Disarmament alleged that CND had been infiltrated by Communists and Trotskyists and was financed in part by the Soviet Union.

Occasionally the smear campaign has backfired on government, as when a letter from the British Home Office sent to its twelve regional scientific advisers was leaked to the press. It gave guidance on the best way of responding to the unsympathetic treatment of the Home Office in the British Medical Association's (1983) report and described the Association as being 'strongly influenced by CND-type propaganda' and its report as showing 'a degree of bias towards the CND case and lack of cogent argument or analysis' (*The Guardian*, 8 July 1983). Whether the smear campaign has worked is another matter. Monsignor Bruce Kent, CND's General Secretary, was reported as saying that it had not fundamentally damaged CND 'but it has had the effect of making it more difficult for people in the professional world to become involved' (Kent, 1983a). Subsequently, commenting on the campaign against CND of British Defence Secretary Michael Heseltine, Kent said: 'It is a pity that he prefers to denigrate us rather than discuss the issues. He is a good showman and propagandist but this sort of thing doesn't help anyone; it just polarises opinion. But at least it shows the government is having to take us very seriously indeed' (Kent, 1983b).

A major feature of the present debate has been the enormous output of books and pamphlets generated both by the peace movement and by those opposed to its views. To some extent the government has contributed to the debate through leaflets and information films supportive of the government case, although many in the peace movement believe that there has been a marked reluctance on the part of the government to debate the issues with CND. Indeed it has been argued by members of the peace movement that the government has hindered open debate of the issues by suppressing information. It is also alleged that civil liberties are being infringed or endangered in an effort to suppress the peace movement.

In spite of this, one of the real achievements of the peace movement has been to bring the issues into the open and ensure that they are debated. In the process the public has become far more informed on defence issues than it was five or six years ago and there is a much greater appreciation, firstly, of the nature of the arms race; secondly, of the size, nature and capabilities of the strategic, intermediate and battlefield nuclear arsenals now deployed; thirdly, of the strategic theories that might govern their use; and fourthly, of the effects that might result from their use. These issues were dealt with in chapters 2 to 7. Two main issues emerge from the debate: the moral argument and the prudential one. There are, as chapter 9 showed, powerful moral arguments against the use of nuclear weapons and even against the threat to use them. There are also good reasons for believing that the manufacture, possession and certainly the use of nuclear weapons are contrary to international law (see chapter 8). Interestingly, the British Government has been anxious to avoid testing the legality of nuclear weapons in the courts. When the first trials of women from the peace camp outside the US cruise missile base at Greenham took place in Newbury, the women were able to raise such issues in their defence. While they were invariably convicted there was, nevertheless, extensive press coverage of their case. Later, steps were taken to prevent protesters from being tried before a jury by dropping certain charges where trial by jury would have been required, or by reducing the charges (*Sanity*, August 1984; *The Guardian*, 23 January 1984). Magistrates have consistently refused to hear defences based on international law.

Moral and legal considerations have often been set aside for reasons of state. This does not mean that actions such as the Soviet invasions of Czechoslovakia and Afghanistan, the secret US bombing of Cambodia to preserve the regime in South Vietnam, the destabilization of Allende's Chile by the USA or its invasion of Grenada, or the detention and sometimes torture of political prisoners in the Soviet Union can in themselves be made ethical, even though they are justified by their perpetrators as necessary in the defence of freedom and Western (capitalist) or socialist (communist) values. They are not ethical, and it is right that those who order them and carry them out should be reminded of this. The fact that the use of nuclear weapons would be immoral and illegal will not deter governments from using them *in extremis*, but it may deter them from using them in lesser circumstances if enough people indicate their opposition on moral and legal grounds. Moreover, the importance of such arguments is that they give a moral and legal basis for opposition to nuclear weapons which it is difficult for governments and those in favour of nuclear weapons to counter.

The prudential argument is, nevertheless, the one more likely to be effective. The contention that Britain's existing defence policy is not credible is based on the following arguments.

1 It commits Britain to a defence policy that relies, *in extremis*, on the use of nuclear weapons, when such use would inevitably bring about the destruction of the country.

2 It allows one of the superpowers to base its nuclear weapons on British territory. Not only would their use result in Soviet retaliation, but Britain has abrogated its sovereignty in as much as it has no control over the use of these weapons by the USA.

3 It commits Britain to an alliance (NATO) whose policy of flexible response is based on first-use of nuclear weapons in a situation in which the Soviet Union has said that it will respond massively and ineluctably to such use.

4 Britain is allied to a superpower that is developing highly accurate, counter-force, war-winning nuclear weapons, and some of whose leaders appear to believe that a nuclear war could be fought and won. Involvement in such a war – and there

can be no doubt that Britain would be involved – would result in the destruction of the country.

If Britain's current defence policy is not credible, the question arises whether other policies might work. There are several possibilities:

1 to adopt a neutralist position outside of an alliance with a superpower, but within a European alliance;
2 to adopt a neutralist position outside of any alliance;
3 to combine 1 or 2 with a nuclear defence policy;
4 to combine 1 or 2 with a non-nuclear defence policy.

It has to be said that *any* defence policy involves a degree of risk. So far as a non-nuclear defence policy is concerned it is argued, for example, that

1 Adoption of a non-nuclear defence policy might encourage the Soviet Union to invade Europe and Britain, thus provoking a highly destructive war. On the other hand, it is argued that this is most unlikely, that a non-nuclear defence policy can act as a powerful deterrent, and that in any case a conventional war would not be *as* destructive as a nuclear war.
2 The Soviet Union might attack a non-nuclear Europe and Britain with nuclear weapons. It is less likely to do so if Britain (and France) have nuclear weapons and if Europe is defended by the US nuclear umbrella. On the other hand, it is argued that having nuclear weapons makes it more likely, if war breaks out, that it will become a nuclear war. It is pointed out that the Soviet Union has said it will not attack non-nuclear states with nuclear weapons. At the very least a non-nuclear defence policy lessens the risk of suffering a nuclear attack.
3 Western Europe could not successfully resist a Soviet invasion without resorting to nuclear weapons. In its extreme formulation this is the 'better dead than red' argument. In response it is argued that the issue is not whether one wishes to live under a communist regime but whether 'the danger from nuclear war is an ultimately greater one to the British national community than the danger from the Soviet Union, and that there is something abhorrent if not lunatic in trying to secure

our cherished values and way of life by the simultaneous threat of quasi-genocide and quasi-national suicide' (Booth, 1983, p. 68).

It is these concerns that recent British governments have failed to address. They have argued instead that possession of nuclear weapons acts as a deterrent and keeps the peace. The evidence for this is based on the fact that there have been 40 years of peace in Europe. It is also suggested that renunciation of nuclear weapons would be a form of appeasement and the parallel is drawn with the situation in the 1930s when the European democracies failed to stand up to the rise of fascism. Apart from the dubious basis of these historical analogies and the unfounded assumption that history will repeat itself, these arguments fail to deal with the central problem of deterrence. As Field Marshal Lord Carver (1982, pp. 101–2) observed:

> It cannot be denied that the concept of nuclear deterrence abounds in illogicalities and paradoxes. At the heart of the problem is the dilemma that if one wishes to deter war by the fear that nuclear weapons will be used, one has to appear to be prepared to use them in any circumstances. But if one does so and the enemy answers back, as he has the capability to do and has clearly said he would, one is very much worse off than if one had not done so, if indeed one is there at all. To pose an unacceptable risk to the enemy automatically poses the same risk to oneself.

It is precisely this paradox that is the focus of the prudential argument. No sane defence policy can lead to a situation in which its execution would result in national annihilation. If the present policy is insane, then it is indeed time to seek an alternative that will serve the needs of the nation in *both* peace and war. As chapter 10 indicated, there are alternatives and it is these that now need to be pursued.

References and Further Reading

Blaker, P. (1983) cited in *The Guardian*, 3 March.

Booth, K. (1983) 'Unilateralism: A Clausewitzian Reform?', in Blake, N. and Pole, K. (eds) (1983) *Dangers of Deterrence*. London: Routledge and Kegan Paul.

Bow Group and Konrad Adenauer Stiftung (1983) *Playing at Peace*, cited in *The Guardian*, 29 March.
British Medical Association (1983) *The Medical Effects of Nuclear War*, Chichester: John Wiley.
Campaign for Defence and Multilateral Disarmament, *Puppets on a String*, London.
Carver, Field Marshal Lord (1982) *A Policy for Peace*, London: Faber and Faber.
Kent, B. (1983a) cited in *The Guardian*, 8 October.
Kent, B. (1983b) cited in *The Guardian*, 4 December.
Nott, J. (1982) cited in *The Guardian*, 15 December.

Acronyms and Abbreviations

ABM	anti-ballistic missile
ALCM	air-launched cruise missile
ASAT	anti-satellite
ASBM	air-to-surface ballistic missile
ASW	anti-submarine warfare
BMD	ballistic missile defence
BMEWS	Ballistic Missile Early Warning System
C	centigrade
C^3I	command, control, communication and intelligence systems
CEP	circular error probable
CIA	Central Intelligence Agency
CND	Campaign for Nuclear Disarmament
CTB[T]	comprehensive test ban [treaty]
DARPA	Defense Advanced Research Project Agency
DEW	Distance Early Warning
EMP	electromagnetic pulse
EMT	equivalent megatonnage
END	European Nuclear Disarmament
ET	emerging technology
FOBS	Fractional Orbital Bombardment Systems
GLCM	ground-launched cruise missiles
GOCO	government-owned, contractor-operated
Gy	gray (absorbed dose of radiation)
HMSO	Her Majesty's Stationery Office
IAEA	International Atomic Energy Authority
ICBM	intercontinental ballistic missile
ICJ	International Court of Justice
INF	intermediate nuclear forces

IRBM	intermediate-range ballistic missle
JCS	Joint Chiefs of Staff
JSTPS	Joint Strategic Target Planning Staff
km	kilometre
LD	lethal dose
LOW	launch on warning
LRTNF	long-range theatre nuclear forces
m	metres [but £m = millions]
MAD	mutual assured destruction
MARV	manoeuverable re-entry vehicle
MIRV	multiple independently-targetable re-entry vehicle
MRV	multiple re-entry vehicle
MUF	material unaccounted for
NATO	North Atlantic Treaty Organisation
NNK	non-nuclear kill
NPT	non-proliferation treaty
NSC	National Security Council
NSDM	National Security Decision Memorandum
NSTL	National Strategic Target List
NWFZ	nuclear-weapon-free zone
PD	presidential directive
PF	protective factor
PGM	precision-guided munitions
PNE[T]	Peaceful Nuclear Explosions [Treaty]
psi	pounds per square inch
PTBT	Partial Test Ban Treaty
R & D	research and development
RAF	Royal Air Force
RV	re-entry vehicle
SAC	Strategic Air Command
SACEUR	Supreme Allied Commander, Europe
SALT	Strategic Arms Limitation Talks
SDI	Strategic Defence Initiative
SIG	stellar inertial guidance
SIOP	Single Integrated Operational Plan
SLBM	submarine-launched ballistic missile
SLCM	sea-launched cruise missiles
SRAM	short-range attack missile
SRBM	short-range ballistic missile
SSBN	ballistic missile-equipped nuclear-powered submarine ['Ship, submersible, ballistic, nuclear']

START	Strategic Arms Reduction Talks
SU	Soviet Union
TERCOM	terrain contour matching
TNF	theatre nuclear forces
TTBT	Threshold Test Ban Treaty
UK	United Kingdom
UKWMO	United Kingdom Warning and Monitoring Organisation
UN	United Nations
US[A]	United States [of America]
USAF	United States Air Force
USSR	Union of Soviet Socialist Republics
WWMCCS	Worldwide Military Command and Control System

Index